Praise for
The Island of Lost Maps

"A magical, endlessly engrossing map of a world most of us probably have thought fusty and uninteresting" —*Chicago Sun-Times*

"Elegantly written . . . Miles Harvey has cleverly tapped into our collective fascination with maps and produced a quirky chronicle of cartographical development disguised as a literary crime story." —*Daily Telegraph*

"An all-consuming read that is impossible to put down."—*Booklist*

"A terrific look into one of the world's million startling subcultures." —*Newsday*

"Harvey's first book [is] as entertaining as the most fancifully illustrated old map you could find." —*San Francisco* magazine

"*The Island of Lost Maps* is a treat. It is part history, part detective story and part journey of self-discovery all rolled into one. The scholarship behind the work is exemplary, and the presentation a joy." —*Fort Worth Star-Telegram*

Island of Lost Maps

BROADWAY BOOKS NEW YORK

MILES HARVEY

The Island of Lost Maps

A TRUE
STORY OF
CARTOGRAPHIC
CRIME

To Bob, Tinker, and Matthew

THE MAPS

To Rengin and Azize

THE DESTINATIONS

BROADWAY

A hardcover edition of this book was originally published in 2000 by Random House. It is here reprinted by arrangement with Random House.

Broadway Books titles may be purchased for business or promotional use or for special sales. For information, please write to: Special Markets Department, Random House, Inc., 1540 Broadway, New York, NY 10036.

BROADWAY BOOKS and its logo, a letter B bisected on the diagonal, are trademarks of Broadway Books, a division of Random House, Inc.

Owing to limitations of space, acknowledgments of permission to quote from unpublished and previously published materials will be found following the Index.

First Broadway Books trade paperback edition published 2001.

Designed by Barbara M. Bachman

The Library of Congress has cataloged the Random House edition as follows:

Harvey, Miles
   The island of lost maps: a true story of cartographic crime / Miles Harvey.
     p. cm.
    1. Libraries—United States—Special collections—Maps—History—20th century. 2. Map thefts—United States—History—20th Century. I. Bland, Gilbert Joseph. II. Title.
   Z702.H37   2000
   025.8'2—dc21   00-025604

ISBN 0-7679-0826-0

10 9 8 7 6 5 4

IT IS NOT DOWN IN ANY MAP;

TRUE PLACES NEVER ARE.

HERMAN MELVILLE,

Moby-Dick

Contents

A PRINT FROM A 1537 EDITION OF A WORK
BY THE THIRTEENTH-CENTURY GEOGRAPHER
JOHANNES DE SACROBOSCO.

Strange Waters

EXPLORERS PIN MAPS TO THEIR WALLS; journalists tape stories to theirs. For both, doing so is a way of getting their bearings. As I sit down to write this book, the wall behind my computer is unadorned except for two photocopied articles, each of which helps me set the course for the journey ahead. The first one, which I ponder now while sipping coffee, is from a reference book called *Who Was Who in World Exploration*:

HOUTMAN, CORNELIUS (Cornelis de Houtman) (ca. 1540–1599). Dutch navigator and trader in the East Indies. Brother of Frederik Houtman. . . . In 1592, the Houtman brothers were commissioned by a group of nine Amsterdam merchants to journey to Portugal to learn what they could about newly developed sea routes to the East Indies. . . . In Lisbon that year, Houtman and his brother attempted to acquire classified Portuguese navigational charts detailing the sailing routes to the Indies. They were arrested and briefly held in a Portuguese jail when they were caught trying to smuggle the charts back to Holland.

The coffee I am drinking comes from a Chicago establishment called the Kopi café, a place where my life was transformed one day and this book was born. As it happens, the story of how the Houtman brothers wound up in jail is also the story of how the Kopi got its name. That, however, is not the primary reason this article hangs on my wall. I keep it there to serve as a constant reminder of the extraordinary power of maps—and the lengths to which human beings will go in obtaining them.

Shakespeare once used the term *mappery* to describe the passionate study of a map or chart. I am neither a map scholar nor a map collector, but if there's one thing I should make clear about myself from the start it's that I am an incorrigible *mapperist,* an ecstatic contemplator of things cartographic. On the desk in front of me now, in fact, is a reproduction of a very old and very famous map, one that, among other things, shows why the Houtman brothers felt compelled to spy on Portugal.

Published in 1569 by Gerard Mercator—the greatest cartographer of his era and a Dutch contemporary of the Houtmans—it was the first map to use a revolutionary system for projecting the three-dimensional world onto a flat plane. More than four hundred years later, this system is still in such common use that when most of us close our eyes and imagine a map of the world, we are seeing Mercator's rendition of the Earth. But for all the work's towering achievements, what I notice at first glance are its mistakes and misconceptions, its whimsies and wild stabs in the dark.

Gazing at Mercator's chart, I see a planet strikingly different from our own, a world full of blank spaces and never-never lands. North America turns into an amorphous blob that reaches so far west it is almost joined to Asia at the hip. South America has an unaccountable protrusion from its southwestern shores, a topographic tail feather that makes the continent look something like a giant waterfowl. This beast is, in turn, perched upon a very peculiar nest—a huge polar landmass, many times the size of present-day Antarctica. Known as the Great Southern Continent or the Unknown Southern Land (or, to more optimistic cartographers, the Country Not Yet Discovered), it was a place ancient geographers had dreamed up to complement their belief that the Earth was perfectly symmetrical. The Arctic region, meanwhile, is a massive donut of land, broken up by four rivers that lead into a polar ocean, through which water was thought to flow to the center of the Earth. From this strange sea rises a giant magnetic "black rock," which, according to ancient versions of the myth, destroyed ships by pulling out the nails that held them together.

It is easy, with centuries of hindsight, to chuckle at such notions. But for people like Mercator and the Houtmans the still-hazy outlines of the world were no laughing matter. Having refrained from widespread ocean exploration until the fifteenth century, Renaissance Europe was in a high-stakes rush to make up for lost time. Portugal had led the charge. Beginning in around 1420 under the direction of a prince who would be dubbed Henry the Navigator by later generations, the Portuguese launched a series of explorations down the west coast of Africa. Bartolomeu Dias arrived at what we now call the Cape of Good Hope in 1488, and ten years later Vasco da Gama became the first European to travel to India by sea. In so doing, he helped establish a Portuguese economic empire in the Far East, the realm of pepper, spices, drugs, pearls, and silk.

The Portuguese controlled the Indies because the Portuguese controlled the maps. Attempting to establish and preserve a trade monopoly, Henry the Navigator and his successors guarded their navigational secrets with an iron hand. "No foreign ship was allowed to sail to the Indies," wrote the historian George Masselman. "The penalty was confiscation of the ship and committal of the crew to a lifetime in the galleys." Nor were maps allowed to circulate. When Pedro Álvares Cabral returned from India in 1501—ending a trip in which he became, according to some scholars, the first European to sight Brazil—an Italian agent complained, "It is impossible to get a chart of the voyage, because the [Portuguese] King has decreed the death penalty for anyone sending one abroad."

On the browned surface of the Mercator chart, I can see signs of just how well Lisbon kept its secrets. Some scholars believe the Portuguese quietly discovered Australia in the early sixteenth century. If so, no word had leaked out to Mercator:

the entire continent is missing from his map. In its place, south of Java, sit three entirely nonexistent kingdoms called Beach, Lucach, and Maletur. Lacking firsthand accounts by contemporary explorers, Mercator based these lands on the fourteenth-century descriptions of Marco Polo.

At first, the Dutch did not consider their ignorance about the Indies to be a pressing issue, primarily because they were close trading partners of the Portuguese. At the time Mercator was making his map, however, that relationship was beginning to change. In 1568 the Dutch, who were increasingly embracing Calvinism, started a long war of independence against Catholic Spain, the great European power at the time. Yet just as the Dutch were breaking from the Spanish empire, Portugal was becoming part of it. In 1580 Spain invaded its neighbor, routing Portuguese forces. Portugal and Holland had now become enemies. The Dutch, brilliant shipbuilders and sailors, realized that they would need to go to the East Indies themselves—no easy task. "The route itself was almost completely unknown," wrote Masselman. "Not a single Dutchman had yet set foot in the Indies."

Their initial attempts to get to the East were by going north. When Dutch merchants looked at world maps such as Mercator's, they saw a waterway passing over Europe and Asia to the Pacific. When they ran their fingers along this route, as I am doing now, they concluded that it was shorter than the passage around Africa—and had the added advantage of being free from heavily armed Portuguese ships. No one had actually made the journey, of course; in fact, the whole concept of this northern waterway—like the notion of the Unknown Southern Land—was based on the writings of ancient historians. It turned out that, for once, the ancient historians were right, but it also turned out that no ship would complete the northeast passage until 1879. The Dutch would make three unsuccessful attempts. In the meantime, they dispatched the Houtmans to Lisbon in search of sea charts. But for all their trouble, the brothers—who were eventually freed after their sponsors paid a considerable ransom—may not even have been particularly successful in their mission. Historians disagree about whether they succeeded in smuggling maps and charts out of the country and, if so, whether these maps provided any new information to Dutch cartographers, who had by this time begun to gain extensive knowledge about Portuguese sea routes from other sources.

What the experts do agree on is that the Houtmans and their associates were "the vanguard of the age of modern capitalism," as Masselman put it. The merchants they represented were in the process of pooling their resources in an innovative enterprise called the Company of the Far East, one of the earliest modern examples of the joint stock company, or corporation. This company was the predecessor of the famous Dutch East India Company, which in turn gave rise to the Amsterdam stock market, the first forum for what the historian Fernand Braudel called "speculation in a totally modern fashion." And as the Dutch East India Company thrived, this same class of merchants would figure out a way to use windmill technology for sawing lumber into the exact shapes and sizes needed in shipbuilding—one of the earliest examples of mass production. The Houtmans were helping to define both the geographic and the economic outlines of the modern world, but this probably never occurred to them. They were just a couple of business-men trying to make a buck.

Today the brothers are mostly remembered as explorers. In 1595, a year after returning from Portugal, Cornelius Houtman served as "chief merchant" on the first Dutch expedition to the East Indies. In 1598 he and his brother led another journey, during which followers of a Sumatran sultan killed Cornelius and im-prisoned Frederik. Over the course of his two-year confinement, Frederik Houtman wrote the first Dutch-Malay dictionary; later, in 1619, he made one of the earliest known sightings of the coast of western Australia. Thanks to efforts of adventurers like the Houtmans and a number of superb cartographers, the Dutch East India Company put together a collection of 180 navigational charts. It had the status of a state secret and was, in fact, called the Secret Atlas. With it the Dutch ousted the Portuguese and be-came the dominant colonial power in the southwestern Pacific.

They controlled the region for more than three hundred years—an entire empire built on getting one's hands on the right maps. By the end of the eighteenth century, that empire's most profitable export was a little bean, which the Dutch had begun to grow in Java in 1696. When brewed this bean pro-duced an exhilarating beverage that the Dutch called *koffie*—a word passed on to the Malay and Indonesian languages as *kopi*.

HICH BRINGS ME TO THE SECOND STORY ON MY WALL, a December 21, 1995, *Chicago Tribune* report about another man who got into trouble for stealing maps:

TAMARAC, FLA.—The small, subdued man in khaki pants asked to visit the rare book room. Library curators checked his credentials, logging him as a visitor from Florida. He went inside.

Minutes later, pandemonium—the man fleeing, security guards chasing, police asking why anyone would steal a map from a 232-year-old library book and sprint through the streets of Baltimore.

Clues, it turns out, rest in Tamarac—home to Gilbert Bland, Jr., alias James Perry, a suspect in the theft . . . at Johns Hopkins University and perhaps scores more [burglaries of old maps] from libraries along the East Coast.

I had no idea, the first time I read those words, that they would soon be chiseled into my mind. Yet I do remember feeling an unusually intense jolt of curiosity, as fiery and bracing as the coffee I had raised to my lips. In those days I spent a great deal of time at the Kopi, a self-proclaimed "traveler's café" whose walls were adorned with masks from Bali and shelves were filled with Lonely Planet guides to far-flung destinations. I was then the literary critic for *Outside* magazine, a great job but one that was beginning to wear on my patience. The books I read were about people who climbed Himalayan peaks or rode bicycles through Africa or sailed wooden boats across the Atlantic or trekked into restricted areas of China. These tales of adventure filled my days and my imagination, yet my own life was anything but adventurous, the hours spent slogging through book after book in a dark corner of that coffeehouse or staring endlessly into a computer screen. The interior of the Kopi was ringed by clocks, each one showing the time in some distant locale, and as I watched the weeks tick away in Timbuktu and Juneau and Goa and Denpasar and Yogyakarta, I began to long for an adventure of my own. Or maybe adventure isn't quite the word. It was not that I had any particular desire to do something death-defying; what I wanted was a quest, a goal, a riddle to solve, a destination. My craving, I believe, was not unlike the one Joseph Conrad described in *Heart of Darkness:*

> Now when I was a little chap I had a passion for maps. I would look for hours at South America, or Africa, or Australia, and lose myself in all the glories of exploration. At that time there were many blank spaces on the earth, and when I saw one that looked particularly inviting on a map (but they all look that) I would put my finger on it and say, "When I grow up I will go there."

Looking back now, I think that what Conrad called "blank spaces" had much to do with why that article so completely commandeered my imagination from the start. In retrospect, I think I was intrigued not so much by what the story said but by what it left looming between the lines. What was it about these mysterious old maps that people found so alluring? And what kind of person would wander so far and put so much on the line for their acquisition? Who *was* this Gilbert Bland? I did not yet know how difficult these questions were. I did not yet know that trying to answer them would take up a huge portion of my life. I did not yet know that, even as I sat sipping coffee, my quest had begun.

All I knew was that I had to know more. I had heard about the bizarre case of Stephen Carrie Blumberg, who, during the 1970s and 1980s, removed as many as 23,600 books and manuscripts from 268 libraries in forty-five states, two Canadian provinces, and the District of Columbia. Consumed by what he called "the passion to collect," Blumberg employed an astonishing variety of tricks to build his illicit collection. He picked locks; he stole keys; he threw volumes out of library windows; he crawled through ductwork and hid in elevator shafts; he assumed the identity of a University of Minnesota professor; he even altered books, while still inside a library, so that they appeared to be his own. Then he would haul his takings to a house in rural Ottumwa, Iowa, where he filled nine thirteen-foot-high rooms with books shelved from floor to ceiling and meticulously arranged them according to a catalog system of his own making. He somehow convinced himself that he was not really stealing but "rescuing" the books from institutions that did not give them the attention and care they deserved. Although his collection was worth up to $20 million, Blumberg never sold a single book. I wondered if Gilbert Bland had a similar compulsion for hoarding hot maps.

But I was just as fascinated by the possibility that Bland was in it for the money. For centuries thieves have reaped big rewards by catering to the peculiar needs of collectors. In our own culture extensive black markets exist for everything from art to animals and sports memorabilia—but history offers some even more outlandish examples. In the twelfth and again in the sixteenth century, for instance, the popularity of a medicinal elixir made from, of all things, the flesh of Egyptian mummies led to a booming business in grave robbing. "Alas, poor Egypt!" wrote Louis Reutter de Rosement. "After having known civilization at its zenith, after having sacrificed its all to respect its dead, it was now forced to see the eternal dwellings of its venerated kings despoiled, profaned, and violated and the bodies of its sons turned into drugs for foreigners."

With our own tedious era sadly devoid of contraband pharaoh goo, Bland's alleged crime spree seemed about as interesting a subject as a writer—especially a writer with his own lifelong love of maps—could hope to find. I began to look into the caper and discovered that it was even more extensive than was originally reported. According to the FBI, Bland had stolen maps from at least seventeen libraries across the United States and two in Canada. He was, it turned out, the Al Capone of cartography, the greatest American map thief in history.

I took the story to my editors at *Outside,* who found the case deliciously offbeat and assigned a lengthy feature on Mr. Bland. I thought I would finish it in six weeks. But when it finally appeared in the June 1997 issue, I had worked on it for more than a year—and my labors had only just begun. By the time I completed research on this book, the investigation had consumed four years of my life.

Bland proved to be an extremely enigmatic and unwilling subject and, despite it all, a fascinating one. He was a chameleon. He changed careers and families without looking back; when a daughter from his first marriage asked him for help with buying a car, he refused, saying, "You're a stranger." He could seem to switch age before your eyes, appearing world-weary one minute and boyish the next. Medium height, medium weight, middle-aged, middle everything—he was a cipher, a blank slate; in cartographic terms, terra incognita. He was Bland: "1. Characterized by a moderate, undisturbing, or tranquil quality. 2. Lacking distinctive character."

Because he turned down all my requests to interview him—both for the article and, later, for this book—I was forced to build my profile the slow, hard way: visiting the places he worked and lived; walking through crime scenes; talking to family members, friends, business associates, and victims; methodically piecing together his past through criminal records, court documents, military files, computer databases, and other sources of public record. The more snooping I did, the more I began to see my quest for information as similar to that of the Houtman brothers, and my attempts to make sense of it like the task faced by Mercator. Filling in a life, it turned out, was like filling in a map, and my search for Gilbert Bland soon transformed from an investigation into an adventure. Along the way,

I happened upon a curious subculture made up of map historians, map librarians, map dealers, and map collectors—all gripped by an obsession both surreal and sublime. Like the explorers of old, I found myself heading farther and farther into strange waters, never quite sure if I had found what I was looking for, but endlessly filled with bemusement and wonder.

THE ISLAND OF LOST MAPS

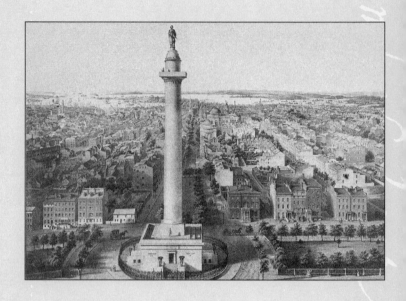

A BIRD'S-EYE VIEW OF BALTIMORE'S
MT. VERNON NEIGHBORHOOD FROM 1850,
AROUND THE TIME GEORGE PEABODY BEGAN
MAKING PLANS FOR A LIBRARY THAT WOULD
ONE DAY STAND JUST LEFT (OR EAST) OF THE
WASHINGTON MONUMENT. ONE HUNDRED AND
FORTY-FIVE YEARS AFTER THIS ENGRAVING
APPEARED, GILBERT BLAND WAS APPRE-
HENDED ON THE FIRST PORCH TO THE RIGHT
OF THE MONUMENT.

CHAPTER 1

Mr. Peabody and Mr. Nobody

THE GEORGE PEABODY LIBRARY IN BALTImore's historic Mount Vernon neighborhood is, by any measure, a remarkable place. "There is no other library like this in America, or anywhere else excepting the parallel universe of Jorge Luis Borges's fiction," the poet and essayist Daniel Mark Epstein wrote. From its lofty reading room, surrounded by giltframed portraits of long-dead librarians, to its Grand Stack Room—measuring sixty-one feet from the white marble floor to the latticed skylight, appointed in ornate cast iron and gold leaf, and containing 250,000 of the world's rarest and most important volumes—the Peabody has lived up to its original conception as "a cathedral of books."

The man who built the library was also remarkable. Born poor in South Danvers, Massachusetts, in 1795, and receiving no more than a fourth-grade education, George Peabody built a small fortune in the dry-goods business in Baltimore, then moved to London, where he made millions in the financial markets. But Peabody was equally adept at giving money away. In England he built more than forty thousand rent-free units of housing for poor people. In the United States he endowed seven cultural institutes and libraries, including the museum of archaeology and ethnology at Harvard and the natural history museum at Yale. At the end of the Civil War, he also established the Peabody Education Fund with $2 million (about $20 million in today's dollars) to provide schooling for the "destitute children of the Southern States." All told, his gifts totaled more than $7 million—something like $70 million in today's dollars.

In 1869, the year of the famous financier's death, the *American Annual Cyclopaedia* called Peabody "the most liberal philanthropist of ancient or modern time"—an overzealous description, maybe, but not by much. Mr. Peabody, as people around the library still respectfully refer to him, is widely acknowledged as the founder of modern philanthropy. He was extraordinary not only in the size of his gifts but in his philosophy of giving. Unlike most big-money benefactors of his day, Peabody did not promote religious beliefs. He had a different sort of evangelism in mind. In an 1831 letter to his nephew, he wrote:

> Deprived as I was, of the opportunity of obtaining anything more than the most *common education* I am *well* qualified to estimate its value by the *disadvantages* I labour under in the society [in] which my business and

situation in life frequently throws me, and willingly
would I now give *twenty times* the expense of attending
a good education could I possess it, but it is now too late
for *me* to learn and I can only do to those that come
under my care, as I could have wished circumstances
had permitted others to have done by me.

Let others save souls; Peabody was interested in minds. And
so it happened that in 1857, during his first visit to the United
States in almost twenty years, he announced plans for a facility
which, he said, "I hope may become useful towards the im-
provement of the moral and intellectual culture of the inhabitants of Baltimore." The Peabody Institute was, wrote his
biographer Franklin Parker, "perhaps grander in its original design than any previous benefaction in America." Its audacious
goal was to jump-start the city's moribund cultural life on many
fronts at once. The institute would have an art gallery and a lecture series and an academy of music. Perhaps most important,
it would also have a library—but no ordinary one.

Peabody had two major stipulations for his library. First, he
insisted that it would be "for the free use of all persons who
may desire to consult it." But, although open to even the poor-
est and least educated Baltimorean, it would not be a lending library, cluttered with contemporary novels, how-to tracts, and
other popular forms of literature. The volumes, he declared,
would be only "the best works on every subject"—and they
would never leave the building. They would remain in the
stacks, always available "to satisfy the researches of Students

who may be engaged in the pursuit of knowledge not ordinarily attainable in the private libraries of the Country." What Peabody had in mind was a kind of modern-day athenaeum, a repository of wisdom and history that would survive the ages.

"Mr. Peabody's purpose," explained Epstein, "was to gather books that had made history and first recorded the thoughts and passions of humankind, in the belief that a book is something more than a locus of data, paper, glue, and ink; a book has supersensible power, and a great collection draws down a resonance endowing the reader of any book in it with special faculties of understanding."

Bringing such a library together was no small task. But library organizers had two huge advantages. The first was lots of money. Peabody eventually endowed the institute with $1.4 million, a huge sum for the time. The second was the Civil War. It delayed the library's official opening for nearly a decade—a vexation for just about everyone but John Morris, the first librarian. "Taking advantage of the delay, Morris and the Library Committee devised a long-range plan to systematically acquire the world's best books for the collection," wrote Elizabeth Schaaf, archivist and curator for the Peabody Institute. "Their model was no less than the combined catalogues of the finest libraries in Europe and America. The Peabody was a carefully planned collection, the scope of which was contingent neither on funds nor immediate availability in the book market."

Morris and his successor, Nathaniel Holmes Morison, carefully compiled massive desiderata—or wish lists—of books for the library. Publishing these lists in bound form, they sent them out to booksellers, bibliophiles, and librarians all over America and Europe in what surely must have been one of the greatest literary scavenger hunts in history. Ten years after the Peabody Institute opened its doors in 1866—an event attended by a crowd of twenty thousand and covered by newspapers on both sides of the Atlantic—there were more than sixty thousand volumes on the shelves, filling the library's original quarters to capacity and prompting the construction of the architect Edmund Lind's palatial Grand Stack Room.

The Peabody collection eventually totaled more than 250,000 volumes—including rare and vital works in U.S. and English history, Greek and Roman classical literature, romance languages, archaeology and art history, science, architecture, building and mechanical trades, genealogy, and, finally, geography and cartography. This amazing catalog has been well-used. H. L. Mencken sat in a reserved desk at the library when working on his monumental treatise, *The American Language*. The novelist John Dos Passos, whose books include *The 42d Parallel* and *1919*, spent countless hours there, along with thousands of less famous researchers and writers. Among them was Epstein, a poet, playwright, and biographer whose books include *The Boy in the Well, Sister Aimee: The Life of Aimee Semple McPherson, No Vacancies in Hell*, and *Nat King Cole*. In a 1993 essay, Epstein described the "magical library" as providing "the necessary environment for private education, the encounter between a person and a book." The Peabody is, he concluded, "a space where time stands still."

That is true in more ways than one. Because of shifting financial priorities at the Peabody Institute, few books have been added to the library's collection since the early part of this century. Today the library—now run by Johns Hopkins University—is something of a gilded warehouse of old books, a treasure-house of knowledge that goes largely unused. Only a few patrons visit the Peabody Library each day. Most of them are scholars, as has always been the case. Others are tourists, who come to gaze at the library's architecture, not its books. Many of the rest are eccentrics, following some odd fancy of intellect or simply killing time. But a few others, constituting a tiny but ominous minority, have darker reasons for being there.

On December 7, 1995, one such man entered the library. It is safe to say that George Peabody had just such a person in mind when he stipulated that the facility "should be guarded and preserved from abuse." It is also safe to say that the stranger was, in many ways, an opposite of the library's founder. Peabody's face—with its distinctive muttonchop whiskers, patrician nose, and wide-set eyes—is still familiar to many people more than a century after the philanthropist's death. The man who walked into the library that day, however, had built a career on anonymity. Few people who crossed paths with him could later recall the details of his appearance; many could not remember him at all. But Mr. Peabody and Mr. Nobody were opposites in even more fundamental ways. Peabody had spent millions preserving rare volumes for the betterment of all people; the man in the Grand Stack Room had come to gut those volumes for his own illicit financial gain.

The two of them, however, had one important thing in

common: a passion to collect. Like the men who assembled this library more than a century earlier, the stranger had come to the Peabody with a wish list. It was a red notebook, roughly the size of a steno pad, its cover bearing the initials U.S.C. and a picture of a gamecock, the mascot for the University of South Carolina, where the man had apparently paid a recent visit. Inside, in neat, well-spaced cursive, was a list of centuries-old books, most of them atlases. Next to many of the titles someone had scribbled the names of various libraries where the books could be found. Several of the entries were followed by "Peabody Inst." or simply "Peabody."

Now some of those same books were piled on a desk in the Grand Stack Room. Now the mystery man began to leaf through one of them. Now he stopped to examine a page. Now he took out a razor blade and carefully lowered his hand to the aged paper.

ON OCTOBER 11, 1492, CHRISTOPHER COLUMBUS, BE-lieving that he was nearing the shores of Japan, looked out across the ocean and saw a light "like a small candle that rose and lifted up." Ever since, historians have debated about what Columbus saw—or whether he saw anything at all. Was it a campfire on a distant shore? Was it the glow of native animals called Bermuda fireworms, which give off blinking green lights to attract mates? "Or was the mysterious light," wrote the Columbus biographer John Noble Wilford, "only an apparition in the eyes of the wishful mariner?" We will never know. What we do know is that four hours later, a lookout on the *Pinta* spotted white cliffs in the moonlight and shouted, *"Tierra! Tierra!"* Cartography would never be the same.

Like other discoveries, the discovery of a crime often depends on a combination of trained eyes, educated guesses, and good luck. Jennifer Bryan had all those things going for her as she sat in the Grand Stack Room that December day in 1995. But her visit to the Peabody would have been completely unremarkable—indeed, the whole saga documented in this book might never have been told—if not for another factor: "I was bored."

Bryan, who was then completing her Ph.D. in history at the University of Maryland, had come to the library to work on her dissertation. Her research that day involved scanning volume after volume of British legal documents from the fifteenth and sixteenth centuries. It was not a thrilling endeavor, even for a student with a sincere passion for her work, and Bryan soon found herself "just sort of staring into space."

From the skylight a gray winter glow washed down over the room. The air was cool and still and fragrant with the faint perfume of old volumes. The place was silent, save for the hush of the ventilation system, the occasional click of footsteps against the marble floor, and the squeal of the old-fashioned elevator. As Bryan sat absentmindedly, the elevator rose into the five upper floors of stacks, carrying a library staff member to collect books. When the elevator stopped, a light came on in one of the balconies, briefly illuminating shelves crammed with old volumes, then went off again. Then the elevator creaked back down to the main floor. This sound momentarily drew Bryan's attention, and when the elevator door opened she watched as a librarian took an armful of old volumes to a man sitting across the way. He was the only other patron in the room.

There was nothing unusual about his appearance—quite the contrary. A studious man in his mid-forties, wearing a blue blazer and khaki pants, he could have been mistaken for half the scholars who walk through the library's doors. He was unimposing and slight-framed, with a biggish nose and smallish chin, reddish hair and mustache—not the kind of person who normally draws stares. Yet Bryan found her eyes lingering on him as he flipped through the books.

"You know how it goes: this is distracting, so I'll watch it to break up the monotony," Bryan recalled about her fixation on the man. In truth, however, she was accustomed to observing other people at libraries. She had been trained to do so. In addition to her graduate work Bryan was employed as a manuscripts curator at the Maryland Historical Society, located just a few blocks from the Peabody. Her desk there faced a huge window that looked down over the James W. Foster Reading Room, and part of her job was to keep an eye on the patrons. She knew that there was good reason for such vigilance. Over the past few years rare books rooms all over the country had been plagued by thefts of increasingly valuable antiquities. The Library of Congress, for example, had announced in 1993 that thieves and "slashers" had stolen thousands of items from perhaps five hundred books in its collections. That august institution was less than fifty miles down Interstate 95 from the

Peabody: what happened there could happen here. Moreover, Bryan knew that the old saying about judging books by their covers also applied to crooks. Often those responsible for the thefts did not look like criminals; in a number of cases, in fact, they had been professors, even librarians. And although she had no obvious reason to distrust the man, she began to get an unsettling feeling about him. "Maybe I just have a suspicious nature," she said.

He quickly gave her good reason to worry. "I just happened to look up and over in that direction and thought I saw him tear a page out of a book," she remembered. But it happened so fast, and the man did it with such seeming nonchalance, that at first she did not completely trust her eyes. "I thought, well, *now* what do I do? Do I say something, or did I just imagine that?"

Bryan decided to wait—and watch. She did not like what she saw. "He was behaving very oddly," she said. "I mean, he wasn't behaving like a researcher. . . . He was just sitting there, riffling through the pages, so it was obvious that he was looking for illustrations."

By this time, the man had noticed her as well. He kept glancing over his shoulder, flashing her "surreptitious" looks. Bryan is an unassuming young woman whose face does not easily give away her thoughts. But behind a pair of glasses, her large blue eyes ache with intelligence and intensity. The man seemed to grow increasingly flustered under her steady stare. "It was weird," she said, "because he must have thought something was up, but he didn't think enough to run, to get out."

Instead, he stood up, pulled out a card catalog drawer, and

laid it alongside the books, purposely obstructing Bryan's view. This was a fatal mistake. She could doubt herself no longer. With that one false move, the man had given himself away.

HEN MR. PEABODY AND MR. NOBODY CROSSED PATHS a few minutes later, there was no time for formal introductions. The encounter took place just outside the front door of the library, where George Peabody, or at least a bronze statue of him, sits leisurely in an armchair, permanently pondering the city he helped to create. At the time of their meeting, Mr. Peabody, wearing a vested business suit and a self-satisfied look, his legs comfortably crossed, had no special plans. Unfortunately, Mr. Nobody was in a rush. At that precise moment, in fact, he was fleeing security guards.

After Jennifer Bryan had told library officials of her concerns about the man, they quietly contacted security officials. Peabody librarian Carolyn Smith then asked the man to move from the Grand Stack Room to a front area, where she hoped to keep an eye on him without arousing his suspicions. The ploy was apparently successful. Even after being moved, the man requested a 1670 atlas of Africa by the cartographer John Ogilby, although perhaps this was simply an attempt to maintain an air of innocence. At any rate, he did not immediately flee—a bad decision. Within a few moments Donald Pfouts, director of security at the Peabody Institute, had entered the room, joined by two other officers. This time the mystery man decided not to stick around. He grabbed his notebook and walked out the front door, followed by Pfouts and the other security officials.

On the library's steps Pfouts said, "Excuse me, sir . . ."

The man picked up his pace. The officers walked faster to catch up—and the man hastened his gait to keep ahead of them. In a scene that might have come from some odd amalgamation of *The Nutty Professor* and *The Fugitive,* the bookish desperado now led his pursuers on a slow-speed chase through downtown Baltimore. Quickening his strides on Mount Vernon Place, Mr. Nobody passed Mr. Peabody and headed toward another famous city landmark, the 178-foot-tall monument to George Washington. "Great Washington . . . stands high aloft on his towering main-mast in Baltimore," wrote Herman Melville in *Moby-Dick,* "and like one of Hercules' pillars, his column marks that point of human grandeur beyond which few mortals will go."

The mortal now in question did venture beyond the column, but grandeur was far from his fate. With security officials now closing in at a jog, an ignominious capture seemed more likely with each moment. He pushed on, circling yet another statue as he headed west across the broad boulevard of Washington Place. Above him a bronze marquis de Lafayette looked down from his horse with a haughty stare. The general's steed, a study in power and movement with muscles taut, mouth chomping at the bit, neck twisting against the reins, tail flying, seemed to be making mockery of the sluggish chase below. Searching for an escape route on the far side of the boulevard, the man spotted a nineteenth-century mansion that now serves as a wing of the Walters Art Gallery. Approaching the building, he ditched his notebook into a row of shrubbery. Then he climbed a stairway onto the Ionic portico.

"That door doesn't go anywhere," Pfouts warned him. The man now realized he was trapped.

Pfouts spoke again: "I would really like to invite you back to the library, because I think there are some issues here that we have to deal with."

When the officers pulled the red spiral notebook from the bushes, they discovered that Jennifer Bryan's suspicions had been well-founded. Folded into its pages were four two-hundred-year-old maps.

S E A   M O N S T E R S   F R O M   A   1550   E D I T I O N   O F
S E B A S T I A N   M Ü N S T E R ' S   *C O S M O G R A P H I A*.

CHAPTER 2

Imaginary Creatures

SOME OF THEM APPEAR TO HAVE THE TORSO of a unicorn, the paws of a weasel, the tail of an anchovy. Some look like Borneo-sized snapping turtles. Some resemble parrot-shark hybrids: the hellish spawn of Big Bird's one-night stand with Jaws. Some look like an amphibious Lucifer, an Evil One with webbed armpits. Some are huge and menacing aqua-dogs, ready to go fetch Madagascar. Some have fur, some have feathers, some have huge human-looking eyes scattered amid their scales. Some are snacking on entire ships. Some have naked ladies surfing on their backs. At least one is bridled like a horse and jockeyed by the king of Portugal. Another has a large sailing vessel beached on its tail, a group of sailors celebrating mass near its gills, and an altar, complete with candles and a crucifix, straddling its dorsal fin. My favorites come fully loaded: spouts, fangs, beaks, spikes, armor, opposable thumbs, the works.

Sea monsters are everywhere on old maps—loitering off Narragansett Bay and splashing around the Arctic Circle, slithering through the South Pacific and causing nothing but trouble near Tierra del Fuego. But they aren't the only unusual forms of life. Hartmann Schedel's 1493 world map, for example, is accompanied by prints depicting a variety of human monstrosities, from one man whose feet point backward to another with ears as big and floppy as Hefty trash bags. In the accompanying text, Schedel explained that in India some men have dog's heads, talk by barking, eat birds, and wear animal skins, while others have just one eye in the middle of their foreheads and eat only animals. Libya has a breed of people who are male on the right side of their bodies and female on the left. The western part of Ethiopia has people with just one mas-

HUMAN MONSTROSITIES ACCOMPANYING THE WORLD MAP IN HARTMANN SCHEDEL'S 1493 *NUREMBERG CHRONICLE*.

sive foot who can run as fast as wild animals. A land called Eripia has beautiful people with the necks and bills of cranes.

Johannes Ruysch's 1507 map, one of the first to record the discoveries of the New World, shows a pair of islands off the coast of Newfoundland said to be inhabited by evil spirits. In a legend next to these isles, Ruysch—who is believed to have gone on one of the early journeys to North America—offered what appears to be an eyewitness account: "Demons assaulted ships near these islands, which were avoided, but not without peril." He is not the only explorer to have gone out into the great unknown and come back telling great untruths. Christopher Columbus reported that the New World contained "one-eyed men, and others, with the snouts of dogs, who ate men," although he conceded he had not seen them himself. He did, however, claim to have spotted three mermaids on his historic 1492 journey but reported that they "were not as pretty as they are depicted, for somehow in the face they look like men." Likewise, Henry Hudson insisted that, while exploring the Arctic in 1608, his crew saw a mermaid whose "back and breasts were like a woman's, her body as big as ours, her skin very white." Sir Walter Raleigh returned from South America in 1596, repeating tales that the jungles were inhabited not only by the Ewaipanomas, a tribe of headless men whose facial features were located in their chests, but by Amazons, a race of warrior women first described in Greek myth. Antonio Pigafetta, a member of Ferdinand Magellan's historic sixteenth-century journey around the world, later wrote that on a beach in what is now Argentina, members of the mission saw a half-naked giant so tall that human beings stood only to its waist.

The colossal beast, called a *patagón* (big-foot), is pictured on a number of seventeenth-century maps, including works made by the great Dutch cartographers Jodocus Hondius and Willem Janszoon Blaeu.

Of course, Patagonia never had Patagons and Amazonia never had Amazons. The explorers were liars. "It is easy enough to move from reporting [on a journey] to embellishing, adding details—perhaps to hold an audience's attention," wrote Steven Frimmer in *Neverland,* a book about nonexistent places and fictive creatures that have persisted as fact in the human imagination. "The next step must have been inventing. . . . The curious thing is that the stories had to be only moderately convincing, as long as they were fascinating. Distant places held a fascination all their own and human nature did the rest. It made the audiences want to believe in the unconvincing parts."

Maybe those early adventurers half-believed their own lies. As Bartolomé de Las Casas, the sixteenth-century missionary who documented the early period of Spanish New World discovery, once observed: "It is a wonder to see how, when a man greatly desires something and strongly attaches himself to it in his imagination, he has the impression at every moment that whatever he hears and sees argues in favor of that thing." Or maybe the explorers made up tales of sea serpents and demons and giants to scare off the competition. Or maybe they found that no one was interested in factual reports, that the public wanted monsters—just as today's public demands tabloid stories of UFO sightings and celebrity love affairs over hard news. Or maybe they created imaginary antagonists so that they could re-create themselves—and return as stronger, smarter, more heroic men. Or maybe they lied simply for the thrill of getting away with it.

Some of them might not even have bothered to leave home. Over the past few years, for example, some experts have begun to question whether Marco Polo ever actually traveled to the Far East. In her book *Did Marco Polo Go to China?* Frances Wood noted that no Chinese records of Polo can be found, despite his claims to have served as an official emissary for Kublai Khan. She also found it curious that, despite his travels, Polo didn't seem to learn much about Chinese geography and that he failed to mention even the most obvious aspects of Chinese culture, such as tea, bound feet, chopsticks, and the Great Wall. Wood, the head of the Chinese Department at the British Library, concluded that Polo probably based his book on the travel tales of other Italian merchants and on Persian accounts of China. It is "unlikely, even allowing for exaggeration," she wrote, that Polo himself ever made it much farther than Constantinople.

He would not have been alone in fabricating his adventures. One of the most important books of the late Middle Ages was *The Travels of Sir John Mandeville,* supposedly written by an English knight who journeyed to China and many other parts of the known world between 1322 and 1356. Like Polo's book, the *Travels* was hugely influential. Christopher Columbus read it in preparation for his 1492 journey. The English explorer Martin Frobisher took a copy of it with him on his 1576 attempt to find the Northwest Passage. The cartographer Abraham Cresques used it as a source for his monumental Catalan Atlas of 1375.

And Martin Behaim's globe of 1492, considered the oldest surviving European-made terrestrial sphere, includes text that quotes Mandeville extensively and respectfully. Details from the *Travels* were being incorporated into maps well into the sixteenth century.

But although the *Travels* was treated as fact for hundreds of years, many of the things John Mandeville reported seeing and doing now seem preposterous. He described walking through the Vale of Devils, for example, a place literally populated by Satan's minions, from which only "good Christian men . . . firm in faith" emerge alive. And he boasted not only of seeing the Well of Youth but of testing its miraculous waters: "I, John Mandeville, saw this well, and drank of it three times, and so did all my companions. Ever since that time I have felt the better and healthier."

Mandeville also claimed to have witnessed an amazing array of life-forms, including geese with two heads, thirty-foot-tall giants, people "who live just on the smell of a kind of apple; and if they lost that smell, they would die forthwith," ants "as big as dogs" that guard "great hills of gold" from human intruders, and "a kind of fruit as big as gourds," with a surprise inside: "an animal of flesh and blood and bone, like a little lamb without wool. And the people of that land eat the animal, and the fruit too. It is a great marvel."

But by far the most marvelous of Mandeville's inventions was his own life. On close analysis, his entire journey seems to have been fabricated, pasted together from other writers' travel narratives. "He was an unredeemable fraud: not only were his rare moments of accuracy stolen, but even his lies were plagia-rized from others," wrote the literary scholar Stephen Green-blatt. Another critic conjectured that the longest journey Mandeville ever took was to the nearest library.

Even the name John Mandeville seems to have been a fab-rication, leaving open—and probably unanswerable—the question of who wrote the *Travels,* and why. "The abundant identifying marks vanish on approach like mirages, and the extraordinarily ingenious efforts to name the author have failed," observed Greenblatt. "The actual identity, the training, the motives, even the nationality of the person who wrote *Mandeville's Travels* have become, under scholarly scrutiny, quite unclear. . . . Man-deville is radically empty."

ALTHOUGH THE STAFF AT THE PEABODY LIBRARY didn't know it at the time, the man they caught with the maps in his notebook also had a compulsion for creating imaginary creatures. The latest—the one whose face appeared on a fake University of Florida student ID card he had presented at the front desk—was named James Perry. But, according to police and court transcripts, there had been many others: James J. Ed-wards and James Morgan and Jason Pike and Jack Arnett and Richard M. Olinger and John David Rosche and Steven M. Spradling and James Bland.

He was no stranger to libraries. In the 1970s, when he was in his twenties, he had visited them often. But this, apparently, was long before he was interested in maps. According to one source who knew him at the time, he would use the libraries to track down the names of people who had died in childhood. Then he would create new identities, using the birth dates of the deceased.

He would invent these creatures, and then he would figure out ways to get people to send them free money. It worked for a while but then stopped working so well. In September 1973 the San Diego police arrested a man named Jason Michael Pike for grand theft. The charge, as the man would later admit in court, stemmed from the fact that "I applied for a credit card, and I used it to get money under false pretenses." At a bail hearing, however, the man conceded that Jason Michael Pike was just an alias. His true name, he told authorities, was Jack Arnett. But it later turned out that Jack Arnett was no more Jack Arnett than Jason Michael Pike was Jason Michael Pike. An alias on top of an alias—this led to an absurd moment in which the San Diego Municipal Court was transformed into a scene out of *Alice in Wonderland:*

> The Clerk: Jack Arnett, aka Jason Pike. . . .
> The Court: Is Gilbert Bland your true name?
> The Defendant: Yes, sir.
> The Court: You are also known as Jason Michael Pike?
> The Defendant: Yes, sir.
> The Court: Did Mr. Langford [the court-appointed defense lawyer] advise you of your constitutional rights, Mr. Arnett?
> The Defendant: Yes, sir.

> The Court: Mr. Bland, are you also known as Jack Ar-
> nett?
> The Defendant: Yes, sir.
> The Court: But your true name is Gilbert Bland?
> The Defendant: Yes, sir.
> The Court: What is your middle name?
> The Defendant: Anthony.
> The Court: Anthony?
> The Defendant: Yes, sir.

But even Gilbert Anthony Bland was a persona of sorts. The defendant was born Gilbert Lee Joseph Bland, Jr., a name he had ditched many years before walking into that courtroom. He pleaded guilty that day, receiving a sentence of five years' probation, but his urge to create new selves did not stop. Gilbert Anthony Bland soon transmogrified into John David Rosche and Richard M. Olinger and Steven M. Spradling. And like horror-movie monsters stitched together out of other people's corpses, these creatures got their inventor into trouble— this time, serious legal trouble. On December 30, 1975, Bland was arrested in Tampa, Florida, for using those identities to defraud the U.S. government in a scheme to collect unemployment compensation for ex-servicemen. He pleaded guilty in U.S. district court, where he was given a three-year sentence and shipped off to the federal corrections institution in El Reno, Oklahoma. He did not like it there, and when he got out he apparently vowed to begin anew.

He cut ties with his old family, started a new one, got a college degree, found work in the computer business, and began to lead a middle-class existence in suburban Maryland. With this new life came a new identity, and an old one at the same time. Gilbert Anthony Bland was now back to his birth name (minus the "Lee"): Gilbert Joseph Bland, Jr. I think he intended this to be his final creation. I think he wanted to leave the phantasms behind. But somewhere in his head they must have been calling to him, enticing him with easy money, adventure, escape.

Then one day in the early 1990s, they began to catch up with him. According to the FBI, it happened almost by accident. "His story is that he bought a bunch of items that someone had left unclaimed at one of these U-Store-It places," said Special Agent Gray Hill, the Bureau's point man on the case. "Included were a bunch of maps. And someone told him, 'Hey, there might be some value there.'"

LONG AFTER THE FBI AGENT TOLD ME THAT ANECDOTE about Gilbert Bland finding the maps, I was left with a vague feeling of déjà vu. Then one day I realized I had, in fact, heard the same story before—many times. On the desk in front of me, for example, is an open copy of *The Clue in the Embers*, the thirty-fifth installment of the Hardy Boys mystery series, a favorite of generation after generation of American children. In the chapter at hand, Joe and Frank Hardy, along with their "chubby pal" Chet Morton, have just discovered a missing medallion. The chums immediately realize that their new find, when matched with its companion, creates a map:

"It must show the area near the treasure that Luis Valez is looking for," Frank remarked.

"And the opal probably marks the place where the treasure is hidden," Joe added. "Boy, I'd like to find that spot myself!"

"But it's in Texichapi—the land of nowhere," Frank reminded him.

You don't have to read the rest of *The Clue in the Embers* to know what happens next. In about the time it takes chubby Chet to exclaim, "Wowee! What a treasure!" the brothers will locate the land of nowhere and uncover the long-buried Aztec palace there. This outcome is certain not only because the Hardy Boys are predestined to leave no mystery unsolved but also because the book is part of a larger literary tradition. So many adventure stories are set in motion by someone finding a map, in fact, that these narratives have assumed the status of myth, in which the buried treasure lies not only in the earth but deep in our collective unconscious. When I went back to read some of them, I was surprised to find how often these stories— like the real-life drama I was covering—were tales of temptation, exploring the perpetual contention between our good and evil selves, between our rational minds and the creatures that prowl the unexplored regions of our souls.

In the case of the Hardy Boys, of course, evil never stands a chance. Being the very embodiment of upstanding American youth, Joe and Frank refuse to exploit Texichapi's priceless antiquities for their own gain. (Frank: "These treasures are certainly government property! No one must be allowed to steal them. We must notify the Guatemalan government at once." Joe: "We don't expect a reward. We've had a grand time visiting your beautiful country.") In other such stories, however, the moral landscape becomes far more complex. At the heart of these narratives are two opposing icons, one symbolizing desire, the other knowledge. If the buried treasure is the forbidden fruit of these stories, the map is the serpent, prodding us to dream of a place beyond the borders of our innocence, pointing us to it, hissing, *X marks the spot.* Perhaps this explains why our culture uses cartographic and geographic language to express notions of sin and virtue. We speak of a moral compass. We describe good people as following the straight and narrow. We say sinners lost their way, lost their bearings. And in our fables of maps and money, characters are constantly torn between sticking to the path of righteousness and wandering into the wilderness of the soul, populated by all those wild animals.

The most celebrated of these stories, of course, is Robert Louis Stevenson's *Treasure Island.* This 1883 classic is often regarded as a harmless children's tale, but as the Stevenson biographer Frank McLynn pointed out, "There is a disturbing darkness about *Treasure Island* absurdly glossed over by those who continued to treat it as a mere 'boy's book.' . . . One of the most disturbing aspects of *Treasure Island* is its lack of a moral center."

The plot is well known. *Treasure Island* opens with the death of an old mariner, who leaves behind a sea chest. Jim Hawkins, a fatherless boy and the book's narrator, finds a mysterious packet of papers in the trunk and takes it to Squire Trelawney and Dr. Livesey for inspection:

> The doctor opened the seals with great care, and there fell out the map of an island, with latitude, longitude, soundings, names of hills, and bays and inlets, and every particular that would be needed to bring a ship to safe anchorage upon its shores. . . . There were several additions of a later date; but, above all, three crosses of red ink—two on the north part of the island, one in the southwest, and, beside this last, in the same red ink, and in a small, neat hand, very different from the captain's tottery characters, these words:—"Bulk of treasure here."

Jim and the men decide to hire a ship and search for the island. Before their vessel departs, however, the one-legged pirate Long John Silver and his associates secretly join the crew and plot to get the buried gold for themselves. This sets up a bloody battle when the ship arrives at Treasure Island. Jim and his friends are eventually able to overcome the buccaneers and find the treasure—but theirs is not a clear-cut triumph of good over evil.

Silver, for one, is a strangely paradoxical villain—especially given the simplistic moral framework of most children's literature. Both a "monstrous imposter" and a man of his word, a reckless killer and a careful compromiser (who, at one point, saves Jim from being murdered by the other mutineers), he is alternately a "prodigious villain" and a "bland, polite, obsequious seaman." The critic Ian Bell aptly described him as a "figure whose personality swings like a pendulum, whose character metamorphoses almost in an instant."

But *Treasure Island*'s hero is also oddly ambiguous. Despite the gee-whiz tone of his narrative, Jim is no Hardy Boy. In his worldview, love of money is not the root of all evil—far from it. Greed is good, and without consequences. Jim does not worry about where the buried treasure came from, who suffered by its accumulation, or whether he has the right to take possession of it. In fact, he revels in his avarice, providing a sensual description of the loot, one that might as easily have come from Long John Silver:

It was a strange collection, like Billy Bones's hoard for the diversity of coinage, but so much larger and so much more varied that I think I never had more pleasure than in sorting them. English, French, Spanish, Portuguese, Georges, and Louises, doubloons and double guineas and moidores and sequins, the pictures of all the kings of Europe for the last hundred years, strange oriental pieces stamped with what looked like wisps of string or bits of spider's web, round pieces and square pieces, and pieces bored through the middle, as if to wear them around your neck—nearly every variety of money in the world must, I think, have found a place in that collection; and for number, I am sure they were like

autumn leaves, so that my back ached with stooping
and my fingers ached with sorting them out.

A villain capable of good, a hero who "never had more pleasure" than in fondling stolen money: in writing *Treasure Island,*
his first novel, Robert Louis Stevenson was already fascinated
by the notion of the divided self, a theme he would explore far
more thoroughly in *Dr. Jekyll and Mr. Hyde.* It's a subject that has
long captured the imagination of writers, not the least of them
being James Fenimore Cooper, creator of such masterpieces of
early American literature as *The Pioneers, The Last of the Mohi-
cans, The Pathfinder,* and *The Deerslayer.*

Cooper's 1849 novel, *The Sea Lions,* is an allegory about two
men on identical ships traveling the same route. If not quite
Jekyll and Hyde, these sailors are nonetheless doppelgängers—
alter egos embodying two very different kinds of human desire.
As in *Treasure Island,* this tale begins with the death of an old
sailor, who leaves behind a sea chest containing "two old, dirty,
and ragged charts." These are no ordinary maps, however. The
first shows the secret site of a seal-hunting paradise in the
Antarctic Circle; the other, the location of a West Indies island
where pirates have buried a "very considerable amount of treasure." Two schooners—both of them named *The Sea Lion*—set
sail in search of this double jackpot. The first ship is commanded by Roswell Gardiner, an honest, brave, and resourceful
Long Islander who, while diligent about making a profit, is primarily on a spiritual journey, having lost his belief in the divin-
ity of Christ. The second ship is led by Captain Jason Daggett, a

Martha's Vineyard sailor, also courageous and cunning—but whom "no dangers, no toil, no thoughts of future, could divert from a purpose that was colored by gold."

In the Antarctic Circle the ships harvest a fortune in sealskins, but because of misfortunes brought on mostly by Daggett's greed, both become trapped in icy waters. During a horrible winter Daggett perishes, confessing to his symbolic twin: "I'm afraid that I've loved money most too well." His alter ego done away with, Gardiner promptly learns humility, realizes that Christianity admits no "half-way belief," accepts Jesus as his savior—and survives. He recovers the buried treasure on his way home, but it proves to be worth only "a little more than two thousand dollars," much less than expected. And, unlike Jim Hawkins in *Treasure Island,* Gardiner understands that the money does not really belong to him. "Seeing the impossibility of restoring the gold to those from whom it had been forced in the first place," he decides on another course of action:

> The doubloons were distributed among the families of
> those who had lost their lives at Sealer's Land. The
> shares did not amount to much, it is true, but they did
> good, and cheered the hearts of two or three widows
> and dependent sisters.

A decidedly less happy, and less virtuous, fate befalls the split-personality protagonist of *The Treasure of the Sierra Madre,* a famous adventure tale by B. Traven—himself no stranger to imaginary creatures. The author's own life, in fact, was a study in the slippery nature of identity. B. Traven was a pseudonym—but for whom? That question, the travel writer Paul Theroux once observed, is "the greatest literary mystery of [the twentieth] century." Over the years, people have identified Traven as Jack London, a group of writers in Honduras, even the illegitimate son of Kaiser Wilhelm II. The most likely bet seems to be that he was the fugitive German anarchist Ret Marut (a name that itself was merely an alias for a man named Otto Feige, according to some researchers). Yet those obsessed with Traven's "real" identity overlook an important point—one that may offer insight into the life of Gilbert Bland. As the Traven biographer Karl S. Guthke observed: "Here was someone who had apparently realized the fantasy of millions. . . . He had enacted the adventure of vanishing into thin air and being reborn, phoenixlike, as someone else—literally a self-made man."

Not surprisingly, the idea of dual identity filters into *The Treasure of the Sierra Madre,* the 1927 novel that was later made into a film with Humphrey Bogart and Walter Huston. The treasure here is a lost gold mine, located somewhere in the Mexican state of Sonora and rumored to have been a primary source of the Aztec empire's wealth. All traces of the legendary mine disappear for hundreds of years, until the late nineteenth century, when college students on vacation in Arizona come across some old maps in the library of a priest:

One of the maps indicated a mine which was named La Mina Agua Verde. When [the students] asked the father about this, he told them the story of this mine. He admitted that the mine was one of the richest known, but said that it was surely cursed, whether by the ancient owners, the Indians, or by the Almighty he would not venture to say. But cursed it was. Whoever went near this mine was sure to be overtaken by misfortune.

The discovery of the map triggers a number of efforts to find the mine—all of them ending in dissension and bloodshed. Yet this fact only seems to strengthen the legend's allure and entice more men to look for La Mina Agua Verde. One of them is Dobbs, an American drifter in Mexico who is particularly fascinated by one part of the myth: a thing he describes as

that eternal curse on gold which changes the soul of a man in a second. The moment he had said this he knew he had said something that never had been in his mind before. Never before had he had the idea that there was a curse connected with gold. Now he had the feeling that not he himself, but something inside him, the existence of which until now he had no knowledge of, had spoken for him, using his voice. For a while he was rather uneasy, feeling that inside his mind there was a second person whom he had seen or heard for the first time.

Dobbs and two other men head off in search of the mine. They find it—but that "second person" takes stronger possession of Dobbs with each ounce of gold dust they pull from the hills. He becomes paranoid and unquenchably covetous, swindling one partner out of his share of the treasure and shooting the other, leaving him for dead. But before Dobbs can enjoy his wealth, he is attacked by bandits, who cut off his head with a machete. They steal his burros—and, thinking the sacks of gold dust are filled with mere sand, they empty the contents to scatter in the desert winds. The curse of La Mina Agua Verde lives on.

Curses are for legends. We rarely speak of them in nonfiction narratives. Still, I can't help feeling that there was a certain inevitability about Gilbert Bland's story from the very start, as if what would follow was both unavoidable and doomed. The superstitious might call this fate. Bland's lawyers described it as mental illness. Bland himself once hinted it had a lot to do with simple desperation. Whatever the case, it must have already been with him when he came upon those first old maps. As I try to picture that pivotal moment, I see him opening some long-sealed cardboard box to discover strange-looking manuscripts stacked inside. He takes them out into the light, blows the dust away, flattens them on a table, runs his fingers over the coarse paper, admires the craftsmanship of the printing and hand coloring. And, even at this early point, he is beginning to hear the voice inside his head, a whisper from the phantom who would become James Perry. Maybe these maps don't *lead to* treasure, this voice tells him, maybe they *are* treasure. And maybe you should find out where more of it is buried.

Then again, it may not have begun that way at all. It's quite possible that Gilbert Bland lied to the FBI about how he became interested in antique maps. The real story could be entirely different—and perhaps, as with the details of B. Traven's life, or John Mandeville's, the truth will never be known. That's the trouble with people who make up imaginary creatures.

I SUPPOSE ALL OF US HAVE STRANGE VOICES IN OUR heads, quirks of cognition that defy simple explanation. When I was a boy, my parents were convinced I had a sixth sense. My father was a schoolteacher with a lot of vacation time and a mean case of wanderlust, so our family would spend much of the summer exploring the U.S. interstate highway system, a great barn of a travel trailer in tow. It was during these long trips that my supposed gift was discovered. Soon, testing it became a kind of game.

I would be fast asleep in the backseat, drooling on my big brother's shoulder, when my father would turn to shake me awake. "Quick," he would say as I blinked my eyes open. "Which direction are we going?"

"Uh . . . north?"

He would look at my mother with a proud and conspiratorial smirk. "Amazing," he would say, shaking his head. "You're right once again."

It was true. As long as I can remember, I've had an excellent sense of direction. But, though I will never fully understand it,

I suspect my boyhood adventures as the Human Compass had a lot more to do with cartography than with clairvoyance. More than likely I simply made educated guesses about the direction we were headed, based on earlier perusals of the coffee-stained, mildew-fragranced highway maps that were as much a permanent fixture of our Chevy Bel Air station wagon as the steering wheel. I could read maps before I could read books—and that skill, combined with a knack for measuring the sun's angles, probably accounted for the better part of my so-called gift. If I was a prodigy at anything, it was cartographic comprehension.

I was good with maps, plain and simple. But what obsessed me about them was never their scientific utility. I did not look on them as mere tools but as mysterious and almost sentient beings. Maps spoke to me. They still do.

Having said that, I want to stress that I am not normally in the habit of listening to mute objects. Nor am I overlooking the obvious fact that a map is a visual medium. Nonetheless, when I need to think through some daunting problem, personal or professional, I often find myself flipping open an atlas and meditating on some random page. And after a few moments of blank contemplation, I no longer seem to be looking at the map so much as listening to it.

A map does not converse in sentences. Its language is a half-heard murmur, fractured, fitful, nondiscursive, nonlinear. "It is almost as if one had to read from a page where all the words had been assembled in random order: obviously there could be no fixed starting point or sequence of perception," wrote Arthur H. Robinson and Barbara Bartz Petchenik in *The Nature of Maps*. A map has no vocabulary, no lexicon of precise meanings. It communicates in lines, hues, tones, coded symbols, and empty spaces, much like music. Nor does a map have its own voice. It is many-tongued, a chorus reciting centuries of accumulated knowledge in echoed chants. A map provides no answers. It only suggests where to look: *discover this, reexamine that, put one thing in relation to another, orient yourself, begin here* . . . Sometimes a map speaks in terms of physical geography, but just as often it muses on the jagged terrain of the heart, the distant vistas of memory, or the fantastic landscapes of dreams.

I use a map to get from one place to another in my mind in the same way that someone else might use it to get from Omaha to Oskaloosa. It's a peculiar kind of travel, I admit, but I suspect many others embark on similar journeys. Robert Louis Stevenson, for one, believed maps had the power of "infinite, eloquent suggestion." Before he wrote *Treasure Island*, he sketched a map of its imaginary shorelines. Soon, he was practically taking dictation from the thing:

> As I pored upon my map of *Treasure Island*, the future
> characters of the book began to appear there visibly
> among imaginary woods; and their brown faces and
> bright weapons peeped out upon me from unexpected
> quarters, as they passed to and fro, fighting and hunting

treasure, on these few square inches of a flat projection. The next thing I knew, I had some paper before me and was writing out a list of chapters.

So important was this map in the genesis of *Treasure Island,* in fact, that Stevenson was reluctant to take too much credit for writing the book. "The map," he wrote, "was the chief part of my plot." He offered this advice to other writers:

> It is my contention—my superstition, if you like—that he who is faithful to his map, and consults it, and draws from it his inspiration, daily and hourly, gains positive support. . . . The tale has a root there: it grows in that soil; it has a spine of its own behind the words. . . . As he studies [the map], relations will appear that he had not thought upon.

Like Stevenson, I often find myself ruminating on maps before I sit down to write. Often, little more comes of this exercise than a sense of relaxation and focus, a general freeing up of the imagination. But once—just as I was beginning work on this book—a seven-hundred-year-old map suddenly seemed to be speaking about Gilbert Bland. *Mappa mundi:* that's the mellifluous expression for a medieval work of cartography—and, despite appearances, it does not mean "map of the world." It more accurately translates as "napkin of the world," a reference to the fact that *mappae mundi* were often painted on cloth.

During the Middle Ages, in fact, there was no word for "map," either in the everyday languages of Europe or in Latin— evidence of how different the medieval conception of geography was from our own. The *mappae mundi* were intended more to diagram history and anthropology, myth and scripture, dreams and nightmares, than to provide geometrically precise representations of the physical world. Not surprisingly, they can look bizarre to modern eyes.

The *mappa mundi* that spoke of Bland was a reproduction of a brightly colored illustration found in a thirteenth-century book of psalms. The original, housed in the British Library and widely known as the Psalter Map, is less than six inches high and four inches wide—just the right size for a napkin of the world—but a much larger version of it is thought to have hung in the residence of the English king Henry III. At the exact middle of the map, like a bull's-eye on a dart board, is Jerusalem, the "navel of the world" in Christian teaching: "Thus saith the Lord God; This is Jerusalem: I have set it in the midst of the nation and countries that are round about her" (Ezekiel 5:5). From Jerusalem extends the known world—Europe, Africa, and Asia (with the last on top because the map is oriented to the east). The unknown world begins at the margins.

In these borderlands lie the spots that medieval Europeans had never actually seen but wanted to exist, the landmarks of their fondest hopes and darkest fears. At the top edge of the Psalter Map, for example, is the Garden of Eden, complete with the images of Adam and Eve standing by the Tree of Knowledge. On the northeastern border are the nations of Gog and

Magog, from which the dreaded servants of the Antichrist were expected to overrun Christendom on the Judgment Day. (Gog and Magog are surrounded on the map by a huge wall, which, according to legend, was built by Alexander the Great to keep those nations' flesh-eating citizens at bay.) And on the map's southernmost margins, standing side by side like criminals in a police lineup, are representatives of the various monstrous races: the Artibatirae, who walk on all fours; the Cynocephali, who have the heads of dogs; the Epiphagi, whose eyes are on their shoulders; the Maritimi, who have four eyes; the Sciopods, who have one giant foot; the Troglodytes, who live in caves. In the coming centuries the borders of the world would expand, and such monsters would move steadily farther from the civilized center, becoming extinct only when their natural habitat, terra incognita, finally ceased to exist.

In some very real ways, however, they are with us still, those monsters. We've added four new continents to our maps since the Middle Ages, subtracted Gog, Magog, and the Earthly Paradise, firmed up our ideas about sphericality, gravity, and heliocentricity. Yet, as I stared at the Psalter Map that day, I was struck not by its geographical failings but by its psychological accuracy. "Like the earth of a hundred years ago, our mind still has its darkest Africas, its unmapped Borneos and Amazonian basins," wrote Aldous Huxley, author of *Brave New World*, in his 1956 book, *Heaven and Hell*. Huxley called these regions "the antipodes of the mind"—a reference to a mythical landmass found on the southern edges of many *mappae mundi* (though, technically, not on the Psalter Map). Just as the geographical antipodes were thought to be inhabited by physical mutants, Huxley's antipodes of the mind are home to "strange psychological creatures leading an autonomous existence according to the law of their own being." Huxley, who found his own way to these antipodes by ingesting mescaline, insisted that we all have "an Old World of personal consciousness and, beyond a dividing sea, a series of New Worlds." For some, he wrote, seeking out "the mind's far continents" leads to a heaven of visionary experience; for others, a private hell.

I was meditating on such notions when the Psalter Map suddenly seemed to be telling me about Gilbert Bland's life, describing in vivid shades of green and red and blue a series of journeys back and forth between the center and the margins, between a mainstream middle-class existence and a criminal life on the edge, between a core identity and a pack of imaginary beings. Then it occurred to me that the map was also saying something about the kind of person drawn to such desperate excursions: you don't go unless you're either extremely self-delusional or self-destructive, unless you think you can reach an Earthly Paradise or are searching for your own version of Gog and Magog. You have to believe in other things, too: that there are places in the world where you can get lost, places where you can leave your old life behind, places where normal rules don't apply. And although I had never been prone to such adventures myself, I now realized that, in following Bland, I would be somehow traveling on a parallel journey to the margins. It struck me that mine, too, should begin at the center.

I knew that, in the antique maps business, no one is more central than a charismatic figure named W. Graham Arader III—though perhaps calling him the navel of the map world would be taking the metaphor too far. Some of his more cynical contemporaries might pick a different body part.

DETAIL FROM A MAP OF THE
TURKISH EMPIRE IN JOAN BLAEU'S
FAMOUS *ATLAS MAJOR*.

CHAPTER 3

The Map Mogul

*I*S THIS *BORING* OR WHAT?" GRAHAM ARADER growled at me as he entered a sales gallery at the elegant New York offices of Sotheby's, the venerable auction house. "Didn't I warn you? Didn't I say that this thing would be a big bore?"

It was late in the afternoon of a summer day so maniacally hot that even Manhattan seemed to have succumbed to a mood of small-town languor. A sale of rare books and manuscripts had been going on since morning, and now the gallery was less than half full, its chairs in disarray. A few hours earlier the place had been crammed with TV crews and scurrying newspaper reporters, all there to witness the much-publicized sale of some recently discovered love letters between Albert Einstein and a reputed Russian spy. The journalists, however, had gone home disappointed: the top bid on the Einstein correspondence—$180,000—proved to be less than the preestablished minimum price, prompting Sotheby's to take a pass on the sale. Since then a steady stream of literary curiosities had gone on the auction block: Sigmund Freud's note to an associate, begging for book royalties; galley proofs for a snippet of Marcel Proust's *Remembrance of Things Past*; Elvis Presley's epistolary vow that his love for a Memphis TV personality could not "be equaled or surpassed by anyone"; and the final shooting script for the second half of *Gone with the Wind,* complete with a proposal to change "Frankly, my dear, I don't give a damn" to "Frankly, my dear, I just don't care." Either line would have suited my mood at that moment. Coming from the Midwest, where auctions usually involve farm implements, I had expected to find this event completely entertaining if not engrossing. But as the afternoon wore on, I had grown tired of the practiced nonchalance of the clientele, who bid thousands of dollars on rare first editions with all the enthusiasm of someone ordering scrambled eggs, and of the endlessly solicitous patter of the auctioneer, a man with the prim voice of a BBC newsreader and the unctuous charm of an American game show host. By now, the pretensions of the place had begun to feel as stifling as the heat outside.

Arader was right: this *was* boring. I knew, however, that it was unlikely to remain that way now that he was on the scene. W. Graham Arader III has been accused of a lot of things by a lot of people, but no one has ever charged him with being dull. Brash and bombastic, brilliant and bigger than life, he has been the subject of lengthy profiles in such prominent publications as *The New Yorker* and *Smithsonian* magazine. He has also served as the thinly veiled model for the central character of a recent novel and is without question the most recognizable figure in the world of antique maps. When I began my research on that world, I quickly learned two things: (1) almost everybody in the business seemed to cherish the chance to discuss Arader, often in reproachful terms (and usually off the record), and (2) Arader had an equal enthusiasm for talking about himself, almost always in the most laudatory manner. "I'm the biggest map dealer in the twentieth century," he said in one typical moment of bravado. "There's no question about it. I sell $10 million in maps every year. I can pick up the phone and make $10,000 in a single hour. Yes, collecting has made me a very rich man."

And yet, as even Arader's worst critics would probably concede, these were not empty boasts. At the time of the Sotheby's auction, June 1998, I had been keeping in touch with this complex, contradictory man for two years. I was drawn to Arader in part because listening to him talk was like consuming a gallon of espresso in one sitting—an overwhelming sensation, to be sure, and one that could leave a decidedly bitter aftertaste, but also an experience that left you wired and, once the buzz wore off, wanting another cup. And while there was absolutely no indication that Arader and Gilbert Bland had ever met, I realized early on that if I wanted to understand the antique maps business I would first need to know something about its biggest player. Over the past few decades the market for old maps had spread steadily from the esoteric fringes to the mainstream. By Arader's own estimation, "It's straight up since I started collecting in 1971, increasing 5 to 20 percent each year." And no one had played a bigger role in this rise than Arader himself.

I had already talked with Arader many times over the phone and had twice traveled to his main residence—an eighty-six-acre estate in the rolling horse country of Middleburg, Virginia, one of three homes that he and his wife own—to sit in his map-filled den and talk with him. By now I felt that we were pretty much talked out. I had come to Sotheby's to see him in action.

The final event of the day—a sale devoted to natural history books and atlases, Arader's two main sources of wealth—was soon to begin. At long tables on either side of the auctioneer, flawlessly dressed young Sotheby's employees sat ready to take phone bids from clients around the world. Those who would be bidding on the premises were finding their seats. As he strode into the gallery, Arader was accompanied by his teenage daughter, Lilli, and his cell phone. He would dote on both of them in the moments to come, turning from one to the other with terse but lively observations about ongoing events. He took a conspicuously inconspicuous chair at the very back of the room, next to a couple of clients or maybe prospective clients—a pleasant middle-aged man who told me he was a novice collector of botanical prints and his wife, who was passing the time by doing needlework. They were clearly not regular auctiongoers, and when Arader discovered that they did not even have a catalog by which to follow the action, he obligingly scooted off to get them one. A few moments later, as he flipped through its pages, briefing the man on the upcoming items for sale, I saw him stop and jab his finger at one of the images. "It's fabu-lous!" he whispered. "*Fabulous!*"

It's hard not to use exclamation marks when quoting Gra-ham Arader: they seem to reflect not just the way he speaks (you get the feeling he even snores in superlatives) but his whole presence. Although his college sport was squash—he was captain of the Yale team—the forty-nine-year-old Arader has the hard-eyed, hard-boned look of a boxer, and a combative spirit to go with it. When he is making a particularly strong point about something, he tips back his head and sticks out his square jaw, as if expecting a punch from some unseen enemy and relishing the chance to return it. Even his name carries a certain belligerence: it's pronounced not *air-uh-der* but *uh-raider,* as in someone who attacks his target by storm. And that was precisely what he planned to do here at Sotheby's. "I won't start kicking ass until Lot 553," he had told me a few days ear-lier. "That's when I'll really start getting nasty. I'll be buying some things—or else prices are going to go very, very high."

Lot 553—a 1748 book of maps and maritime adventure tales by George Anson—was the very first item up for sale in this final phase of the auction. Its number was now posted on a big electronic tote board, designed to convert any bid instantly into its equivalent in pounds, francs, deutsche marks, lire, yen, and zlotys. As the bidding began, Arader's competitors turned toward the auctioneer's podium. Many of them were dressed in upscale suits, a measure of their respect for the august sur-roundings. Arader himself, I could not help but notice, was heading into battle in a tennis shirt, slightly rumpled green khakis, and a pair of Nikes.

HE FIRST TIME I INTERVIEWED GRAHAM ARADER—I WAS working on the *Outside* magazine story at the time—he used some extremely offensive language to describe a group of people he didn't like. One of the things I greatly admire about Arader is his willingness to speak his mind on almost any subject at almost any time, but these particular words were so objectionable, not to mention bizarre and beside the point, that I probably would not have been able to work them into the article even if I had wanted to. That, however, was not what I told Arader when he called me some time later, asking me to pull the quote. I simply reminded him that our entire conversation had been on the record and that, like any journalist, I could not allow a source to dictate what I wrote or didn't write. At first he began to lose his temper—a legendary trait—but then he calmed down and made me a simple offer: What if he gave me an even more quotable quote on the same general subject? Let me hear it, I said. He did just that. It was indeed more quotable—irresistibly so—and I ended up using it in the article, feeling that I had been very clever in my dealings with him. But months later, when I repeated this anecdote to Hugh Kennedy, a longtime Arader observer, he began to laugh. "See?" Kennedy said. "He's even got *you* buying and selling."

Kennedy knows plenty about such matters, having studied the controversial businessman for years, from the perspective of both employee and author. He worked for Arader from 1987 to 1990, starting out as a gallery assistant and later becoming the director of the Philadelphia office. (Arader also has galleries in Houston, San Francisco, and New York, as well as a kind of up-scale annex, an ornate Beaux Arts town house on Madison Avenue, which he modestly describes as "the single most beautiful house in New York City, without question.") After leaving Arader's service, Kennedy began a career as a novelist. His second book, *Original Color,* published in 1996, tells the story of a high-powered map and print dealer who "pitched fine art with the dogged persistence of a used-car salesman." Kennedy's Arader-inspired character terrifies employees, bends laws, lies to customers, bounces checks, and farts in public—not an entirely complimentary characterization, in short. Nonetheless, it was Graham Arader himself who had insisted that I read the book.

I asked Kennedy why Arader would do that. "Because, in some ways, the book is about him," he replied. "And he is, to a certain extent, self-obsessed. On the other hand, I've met very few people with his charisma. He's completely driven, and can do business any time of the day or night. He also blurs the lines with his clients between vendor-buyer and friend. In any market where you're selling something of premium value, you often need to sell yourself as well as the product. And the entertainment value of Graham is huge. When he's on, he's an enormously fun person to be around, and I think a lot of his clients have stuck with him just for that. They're amused and occasionally delighted by him."

I shared those feelings. Arader could be a shameless flatterer and an equally shameless bully, but there was something very arresting about him nonetheless—not charm, exactly, but a kind of infectious exuberance for everything he said, did, thought, bought, or sold. And somehow, underneath all his seeming self-absorption, he had a skill for listening, for feeling out the needs of others and telling them exactly what they wanted to hear. Even a skeptic like Kennedy was not immune to his almost magical powers of persuasion. Once, on a book tour visit to Philadelphia, Kennedy stopped by to say hello to his former employer. "Within five minutes," Kennedy remembered, "he was offering me a hundred-thousand-dollar-a-year job in the New York gallery, doing sales. I just felt like, 'If I don't get out of this room in five minutes, I'm going to say yes, and I won't even be able to help myself.' There was something bizarre about it."

Arader has used his uncanny charisma to amass a fortune worth, by his own count, $100 million. In doing so, he has led a dramatic transformation in the market for antique maps over the past quarter century, turning a historical artifact into a hip commodity. Before he entered the business in the early 1970s, old maps were mostly the province of librarians, historians, and a few tweedy collectors. One of them was Arader's father, Walter G. Arader, a successful banker and businessman who began collecting maps in the mid-1960s—a passion that stemmed from his days as a navigator in the Navy. Walter Arader typically paid one or two hundred dollars for individual maps from atlases by great Age of Discovery cartographers such as Willem Janszoon Blaeu, Gerard Mercator, and Abraham Ortelius. Today those same maps are listed in Graham Arader's catalogs for prices ranging from five thousand to fifteen thousand dollars.

The younger Arader got his feet wet in collecting as a teenager, during a year he spent in England before college. While studying at the Canford School in Dorset, he made frequent visits to map and book shops in London, where he carried out buying assignments for his father. It was during this time that he began to become enamored of maps himself. "The first map that I remember being wildly excited about was a map by Christopher Saxton of Dorset, which was printed in London in 1579," Arader remembered. "It had beautiful late Renaissance designs on it and wonderful color. And it also showed where I went to school. Like most people, my interest in maps was very provincial. It was just incredibly exciting for me to buy."

But if Walter Arader saw maps as a fun way to *spend* his riches, his son viewed them as a potential way to *make* his own. When he enrolled at Yale, Graham Arader started to turn his enthusiasm for maps into an enterprise. The school has two great cartographic collections, the Beinecke Rare Book and Manuscript Library and the map room at the Sterling Memorial Library. Together, they house literally hundreds of thousands of rare maps and atlases, including the controversial Vinland Map, reputed (by a dwindling number of experts) to be the only pre-Columbian map showing Norse discoveries in America; groundbreaking New World maps by Henricus Martellus and

Johannes Ruysch; and a one-of-a-kind atlas previously owned by George Washington. The young Arader immediately sought these libraries out, determined to give himself a crash course in cartographic history. The scholars who worked in those secluded haunts were stunned to find a college freshman in their midst, but, as Arader would later explain: "That was my advantage. There were thirty world-class rare book librarians, historians, curators of collections—and nobody had any interest in what they were doing. So here were these lonely men and women, without anyone exhibiting any interest in their work, and then here comes this bright, young, aggressive man deluging them with questions. I'm the only guy in history who used Yale as a trade school."

He was still living in a dormitory when he began his career as a dealer. His initial break came when he cajoled a New Haven bookseller into giving him the names of three members of the Yale medical faculty who had paid to have maps framed at the store: "That's how I got my first customers. I called them up and said, 'Are you interested in collecting maps?' They said they were—and they're still good customers to this day."

In 1973, using a $150,000 loan from his father, Arader set up shop in earnest, issuing his first catalog. Business boomed. "It was incredible," he said of those early days. "You could buy the things very cheaply. You'd buy something for a hundred bucks and sell it for two hundred. You know, you doubled your money pretty easily back then."

But why settle for only two hundred dollars, when you might get a whole lot more for a given map? Arader understood that antique maps prices were low because the market was tiny, and he was determined to expand the base of customers beyond the insular realm of aficionados. Some of his competitors probably had similar aspirations, but few of them had his capital, creativity, or connections—and none of them had his charisma. That was the important thing. Arader realized that through his own hustle, savvy, and force of personality, he could actually *create value.* An experienced collector might not be willing to pay three thousand dollars for a map that Arader had purchased for only a thousand. But a wealthy physician, looking for something interesting and fashionable to hang in his waiting room, might indeed be willing to spend that much—especially on a map that came with the imprimatur of W. Graham Arader III, a dealer known to carry only the finest merchandise and devote his considerable energy, talent, and knowledge to only the most exclusive clients. And eventually—after Arader was able to make a few more such sales, *and* after other dealers adjusted their prices accordingly—the map was no longer worth a thousand dollars in anyone's mind. It was now a three-thousand-dollar map—that is, until Arader decided to ask nine thousand for it.

By reaching out to nonexperts with cash to burn and corporations with offices to decorate, Arader has been able to steadily increase the demand for old maps. In so doing he has made many of his clients, not to mention some of his competitors, considerably more wealthy. He also has helped give antique maps unprecedented visibility, not only as investments ("These Old Maps Offer You a New Way to Double Your

Money," read one optimistic headline from the March 1997 issue of *Money* magazine) but as mass-media artifacts. You can now find images of historic maps on greeting cards and wrapping paper, T-shirts and photo albums, coffee mugs and clockfaces, popcorn canisters and computer mouse pads. You can see them in casinos and coffeehouses, children's books and *Playboy* magazine spreads. They have quietly become pop culture icons.

Arader has also helped transform other markets—most notably, the one for prints by John James Audubon and other natural history artists. These days, the biggest share of his business comes from such works. Antique maps total only about 20 percent of his trade. Nonetheless, he insists that "maps are my first love and always will be my first love."

Love, however, is a complicated thing. For all the time and passion Graham Arader has poured into the world of antique maps, the world of antique maps has shown him precious little gratitude. Respect, maybe, but not gratitude. And certainly not love.

*O*NE WAY OF MAKING A BID AT SOTHEBY'S IS TO TAKE THE little paddle provided by the auction house specifically for that purpose and raise it into the air. This is the method of the inexperienced and unhip. Those in the know rely on subtler techniques: the pointedly flicked wrist, for instance, or the pithily raised finger. Graham Arader doesn't use his hands at all. He bids with his head. A quick nod, accompanied by a fiery stare over his reading glasses, indicates that he is willing to top the previous bid; a side-to-side shake, so slight he seems to be experiencing nothing more than a sudden draft, means that he is bowing out. As the auction progressed that afternoon, Arader's skull was in a constant state of jiggle. He looked like a very tightly sprung version of one of those Bobbin' Heads dolls.

He did not purchase Lot 553, the first work up for bid, but he did buy the next two items—a nineteenth-century book of bird engravings by Jean Baptiste Audebert and Louis-Jean Pierre Vieillot for $7,500; and a first edition of John James Audubon and John Bachman's *The Viviparous Quadrupeds of North America* for $200,000. Arader was just getting started. In a flurry of spending over the next few minutes, he would snap up Mathew Carey's *Carey's American Atlas* for $2,000; Mark Catesby's *The Natural History of Carolina, Florida and the Bahama Islands* for $100,000; and John Gould's *A Monograph of the Trochilidae, or Family of Humming-Birds,* again for $100,000. On a number of other items—including natural history books by George Edwards and Gould; lithographs of Native Americans by George Catlin; and travelogues by Captain James Cook—he would help to drive the bidding up, often pulling out at the very last opportunity.

On Lot 585—the geologist and mapmaker Ferdinand V. Hayden's 1876 book on Yellowstone National Park—Arader won the bidding, then immediately reached for his cell phone. "You got it. One hundred and fifteen thousand," he barked to a client whom he was representing at the auction (and whom he later identified as Mark Rockefeller, Nelson's son). "Congratulations."

As he was finishing the call, another man strolled by him on his way out of the gallery. "Quiet, Graham," the man said, apparently in jest.

"You want me to be quiet? All right." I couldn't tell whether Arader had taken it as a friendly gibe or an insult. Whatever the case, his voice was noticeably louder when he added, ever so slowly, *"I'll be as quiet as you please."*

It was obvious that he did not consider the other bidders here his friends. He was not above slapping a few backs and shaking a few hands, but, in general, he kept his distance from them, and they kept their distance from him. (One of the other participants, who had noticed me taking notes on Arader, later pulled me aside. "That guy's crazy, you know," he said. "Did he use any foul language when you talked to him?") Taking all this in, I was reminded of a conversation I had had with Arader a few weeks earlier.

"What about the claim that you're abrasive?" I had asked.

"I *am* abrasive!" he replied. "I don't have time to be nice to my competitors. So I agree with that: I'm abrasive."

"Ruthless? Cutthroat?"

"I'm not ruthless. I'm a businessman. Tell the people who call me ruthless to get into the bond market or the stock market. You want to see ruthless? Tell them to try to compete against Bill Gates selling software. I'm a cupcake compared to real ruthlessness. I will buy whatever I can that's of high quality. So I guess they call it ruthless when I buy all the good things that come up at auction. They end up going home with nothing; I end up going home with everything. Is that ruthless? No. It's business."

I could not deny it: this pretty well described what was happening here at Sotheby's. But I also knew that while simple jealousy did explain some critics' resentment of Arader, others had real philosophical differences with him, not to mention longstanding concerns about his ethics. In 1983, for example, the Antiquarian Booksellers' Association of America drummed him from its ranks. The ouster stemmed from a feud between Arader and a Dutch bookseller who had sent him a five-volume set of John Gould hummingbirds (similar to the one he purchased today at Sotheby's for a cool hundred grand). Arader had decided that the books were in unsatisfactory condition—and, in a typical moment of pugnacity, returned them to Amsterdam third-class and uninsured. The genteel governors of the ABAA found this slight unpardonable. They booted Arader—after which he upped the stakes by filing suit against the group. He lost, and wound up having to pay the ABAA's legal fees as well as his own, to the tune of fifty thousand dollars.

If the incident cemented Arader's bad-boy reputation—an image he clearly relishes—it also added to concerns about his business practices. And of those practices few have been as controversial as a tactic he made famous early in his career. It is known by a number of euphemisms: book breaking, disbinding, discollating, and liberating from bound portfolio. Purists simply call it destroying books.

Graham Arader never claimed to be a purist, however, and when he was trying to make a name for himself in the 1970s, the sum value of the individual maps from an atlas was often greater than the value of the atlas itself. Predictably, Arader acted as businessman, not as bibliophile: he would buy up atlases and other art books, cut them apart, and sell off the plates one by one. He did not invent the concept—in fact, people have been breaking books for centuries—but he put it into practice so methodically and on such a large scale that, to this day, some lovers of literature still refuse to forgive him.

True, Arader's competitors might break an occasional atlas, especially if it already had missing plates or was otherwise damaged. But Arader systematically purchased pristine volumes with the express purpose of demolishing them. In one such instance, a fellow dealer named Bill Reese watched him break a first edition of John Smith's 1624 work, *Generall Historie of Virginia, New-England, and the Summer Isles*. Reese later told *The New Yorker* how Arader laid the book on a counter and began to massage its "gutter," or inner margins. "I said, 'Graham, what are you doing?' He said, 'What do you think I'm doing, Billy? I'm testing the strength of the binding.' Then—r-r-r-ip!—he razors the son of a bitch out."

It sounds ugly. Yet the more I looked into the matter, the more I wondered whether Arader really deserves to go down in infamy as the Bogeyman of Book Breaking. A great deal of hypocrisy surrounds this issue, as I once learned at a map fair where a California dealer named Robert Ross hosted a seminar on collecting. Someone asked Ross about the dreaded practice of cutting. He responded by telling a story about a time when he himself almost broke a low-quality atlas. "I had the knife out ready to go, but I couldn't do it," he said. "I put the knife back, because the atlas was complete."

But if Ross was hesitant to slice up atlases with his own hands, he seemed more than willing to let others do the dirty work. Most of the items hanging on the walls of his booth at the fair appeared to be from broken books, and the same was true for virtually every other dealer at the event. The reason was simple: precious few stand-alone antique maps are in circulation. Without broken books, the antique maps trade would virtually cease to exist.

So many atlases have been broken, in fact, the practice is becoming increasingly unprofitable—even for Graham Arader. "I don't break books anymore. I don't have to," he said. "Books are worth more than the sum of the parts at this point. Is it how I made my money in the first place? You bet your ass it is. Do I feel badly about it? Yes, I do feel badly about it. Looking back now from a very secure and lofty perch, I can say I wish I hadn't been able to do it. But I couldn't have made it without doing it. I mean, you could buy one of these atlases for ten thousand dollars and break it for a hundred thousand. It was wild. Now you buy one of these books for a hundred and fifty thousand and it breaks for ninety thousand. You just don't do it."

He insists he has not broken "a complete, perfect atlas" in more than ten years. And yet questions about his ethics linger—in part, because he keeps getting embroiled in new controversies. In 1996, for example, Arader reportedly sent a check for

$8,000 to a woman as payment for a 160-year-old map, believed to be the earliest plan of the city of Houston. He then had the piece restored and, in a quintessential application of Araderian economics, put it back on the market for $98,000. ("Yeah, I tried to buy it as cheaply as possible," he later told me. "Yeah, I low-balled it. . . . I paid as little as I could for it, to put it bluntly. Who am I going to take advantage of the most: the people I buy from or the people I sell to? . . . Because, boy, I'll sure as hell lose a client if I sell him something at more than it would go for at auction.") At that point the story took a strange turn. A Houston man named John Fox filed suit against Arader, claiming that the map rightfully belonged to him. Fox said that his sister-in-law had taken the map to Arader, and that she had done so to have it appraised, not to sell it. The suit was settled in Fox's favor: Arader agreed to ship it back and pay court costs. But then the story took a second curious twist. Employees of the Houston Title Company, who had read about the legal battle in a local newspaper, began to wonder whether the map was the same one that had disappeared some years earlier from their office, where Fox had once worked. It turned out that their suspicions were well-grounded. After confidential negotiations, Houston Title got back its map, and Fox (who claimed he had purchased the map from the company) never faced criminal charges.

For his part, Arader cooperated with police to ensure that the map was returned to its rightful owner—just as he had done in a number of similar cases. "To this day," he later said, "I am furious that they did not prosecute Fox. . . . The whole bad

taste in my mouth from the thing is that the FBI and local police chose to let the case drop because the people at Houston
Title wanted to sweep the thing under the rug, since they were
embarrassed about it." Yet the incident also revived old questions about the dealer himself. Had he dealt honestly with the
woman who brought him the map? Did he have clear ownership when he put it up for sale? Had he been aware that only
one other copy of the map—a later version at the Houston
Public Library—was known to exist? And, if so, had he done
enough to check its provenance?

These are all, of course, important concerns. But what people will remember most about the incident, I suspect, is that
whopping ninety-thousand-dollar markup on the map, that in-
your-face 1,225 percent price hike. It may not be fair to hold
Graham Arader entirely responsible for inflation in the antique
maps trade. Yet as I watched him at Sotheby's that day, buying
up many items and bidding up others, it occurred to me that
nothing—not his egotism or abrasiveness, not his book breaking or business imbroglios—has earned him more enemies. Yes,
the changing market has spawned some big winners, not the
least of them being Arader himself. But it has also created a lot
of losers, among them middle-class collectors who no longer
have access to the kinds of maps they once could easily afford,
and small-scale dealers who find themselves without the capital
to maintain adequate inventory. These are the same people
who used to form the backbone of the industry—scholars and
aficionados who care about maps primarily for their historic
and aesthetic, not economic, value. They may have helped kick

Arader out of the ABAA—but in a more profound way he not only forced his way into their refined little world but took it over. It's no wonder they regard him with contempt.

Arader seems well aware of this. I once asked him: "Has map collecting changed much since you began?"

"Yeah. You've got to be richer," he said. "Stuff has gone up faster than inflation. So a little librarian in Urbana, Illinois, who ten years ago was buying twenty maps with his twenty-thousand-dollar-per-year budget, is now buying one map."

True enough. But for librarians in particular that's not the worst thing about rising prices. The worst thing about rising prices is that they have transformed libraries into gold mines, books into illicit booty, and patrons into crooks. If the librarians hate Arader, it's not because he's made it harder for them to buy books but because they're having one hell of a time holding on to the ones already on their shelves.

DRAMATIC EXAMPLE OF THE WAY IN WHICH OLD MAPS are becoming big business was about to play out at Sotheby's. Lot 599 was an edition of the ancient scholar Ptolemy's famous map book, *Geographia* (also known as *Cosmographia*), printed in Ulm in 1482. If you wanted to buy this volume in 1884, it would have set you back $85. It went for $350 in 1901, $3,000 in 1933, $5,000 in 1950, $28,000 in 1965, and $42,500 in 1984. Today, the catalog listed an estimated price range of between $200,000 and $300,000.

"On the other hand, it is an extraordinarily fine copy, it's in a contemporary binding, and it has already received a great deal of interest," Selby Kiffer, a Sotheby's official, had told me a few days earlier. "One never knows what will happen at an auction, but I do think that this is a case where competitive bidding could push price beyond estimate range."

From the start, it was clear that Kiffer's hunch was right. The bidding began at $100,000 but quickly moved past $150,000, then $200,000, then $250,000. A sense of urgency filled the gallery, as bids came in quick succession, from the floor and over the phone.

For once not even Graham Arader was willing to keep pace with the competition. He dropped out around $300,000.

The bidding continued to rise . . .

It was hardly the first time buyers had clamored to get their hands on a copy of *Geographia*. The work, in fact, holds a central place in the history of collecting, as well as the history of cartography. Its author, Claudius Ptolemy, was one of the ancient world's most intriguing and shadowy figures. Little is known about his life, other than that he lived in Alexandria during the second century A.D., was of Greek descent and was a Roman citizen. Nonetheless, he left behind a brilliant and wide-ranging body of work about astronomy, optics, mathematics, music, and geography. *Geographia*—a kind of empirical how-to guide for map drawing—is considered the high-water mark of ancient earth knowledge. Ptolemy systematized cartography by insisting that maps be drawn to scale and that they be oriented to the north. He was one of the first to offer a projection by which a spherical earth could be rendered on a flat surface.

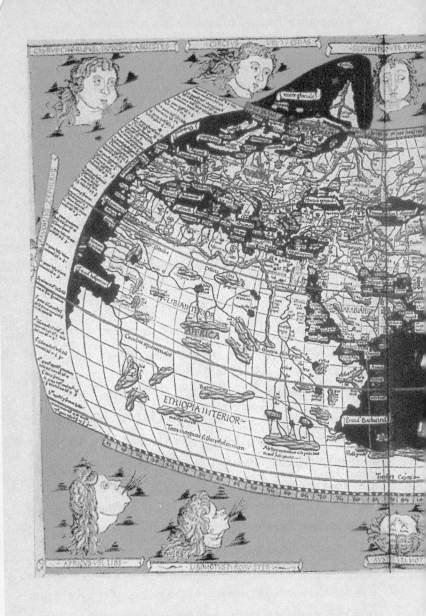

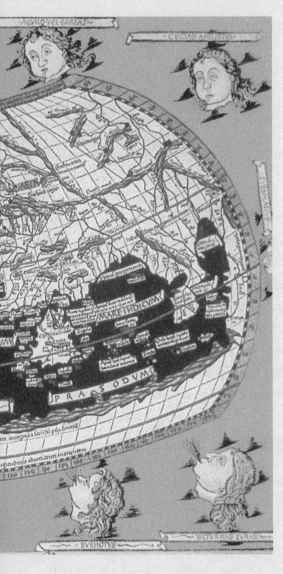

Island of Lost Maps

He also abandoned the Homeric conception that the known world (Europe, Asia, and Africa) was surrounded by an uninhabitable ocean. This left open the theoretical possibility of further discoveries. "More than any one of the ancients," concluded the map historian Lloyd A. Brown, "Claudius Ptolemy succeeded in establishing the elements and form of scientific cartography."

But *Geographia* might not have been remembered that way if not for the efforts of some passionate collectors more than a thousand years later. It might not have been remembered at all. Like many great works of antiquity, *Geographia* simply disappeared from the European consciousness after the fall of Rome. Its concepts, meanwhile, were kept alive by Arab geographers, who translated Ptolemy's cartographic masterpiece around the ninth century and incorporated its concepts into their own maps. Then, in the thirteenth century, a Byzantine scholar and monk named Maximus Planudes found a long-forgotten copy of the work. According to Planudes' account of the discovery, the manuscript was not accompanied by maps. (Indeed, modern scholars doubt whether Ptolemy ever included maps with the work.) Nonetheless, Planudes set about drafting a series of maps designed to portray the world as Ptolemy would have drawn it himself. This was possible because Ptolemy had made the effort to include the geographical coordinates of eight thousand places throughout the world, so that someone in another place—even another century—could create maps on the basis of the text alone. I like to imagine Planudes in his little chamber, plotting coordinate after painstaking coordinate, then at last stepping back from his work, the whole world suddenly appearing before his eyes.

. . . $325,000.

$350,000.

$375,000.

$400,000 . . .

Planudes was an omnivorous collector of ancient manuscripts, who could often be found scouring the bazaars of Constantinople for the works of great classical writers. In addition to translating many works from Latin into Greek, he put together a number of anthologies of lasting importance, including a compilation of Greek prose and poetry, a volume of Aesop's fables, and the marvelously titled *Very Useful Collection Gathered from Various Books*. In his determined pursuit of classical texts, Planudes was a spiritual and intellectual forerunner of the great scholar-collectors of the fourteenth, fifteenth, and sixteenth centuries, who, through "grinding persistence . . . in the recovery, collation, criticism, and publication of texts," transformed "the study of the ancient world into a cultural force," wrote the historian John Hale in *The Civilization of Europe in the Renaissance*. Not the least of these men was the Italian poet Petrarch, famous for his collection of—and obsession with—ancient texts. "Please, if you love me," Petrarch wrote to a friend sometime around the year 1346, "find people who are educated and trustworthy and set them to scour Tuscany, to turn out the book-cases of the monks and all the other scholars, and see if anything comes to light which will serve to quench—or, shall I say, increase—my thirst."

. . . $425,000.

$450,000.

$475,000.

$500,000 . . .

Petrarch confessed that his urge to obtain books was an "insatiable desire which I so far have been quite unable to control." Such compulsions would soon be the norm, as the Renaissance gave rise to an unprecedented culture of collecting. No longer did people seek out art and artifacts solely for their devotional purposes but for their intellectual, historical, scientific, aesthetic, nostalgic, or commercial significance. This collecting ethos—as exemplified by the "cabinets of curiosities" placed prominently in homes of the wealthy—began with an interest in ancient books, gems, coins, vases, and sculpture, then grew to encompass contemporary paintings, clocks, and scientific instruments, and, finally, expanded into what Hale described as "rare, valuable, or merely strange objects from the natural world," from shells and fossils to stuffed toucans and mummified Egyptian cats.

. . . $525,000.

$550,000.

$575,000.

$600,000 . . .

Geographia was at the center of this collecting craze. Around 1400 a copy of the text was brought from Constantinople to Florence, where, translated from Greek into Latin, it "caused an immediate and enormous stir," wrote Thomas Goldstein in *Dawn of Modern Science*. Hand-copied, hand-illustrated versions quickly began to circulate in Western Europe, usually with maps based on those of Planudes. As Lisa Jardine observed in *Worldly Goods: A New History of the Renaissance:*

> Apart from the extravagant Bibles, the ancient scientist and cartographer Ptolemy's *Geography,* complete with coloured and illuminated maps of the known world, took pride of place in a surprisingly large number of great men's libraries. The *Geography,* too, was an extremely expensive purchase, since some copies contained as many as sixty individual maps, each of which had to be accurately drawn and locations precisely marked before the delicate business of colouring and decorating could even be begun.

The first printed edition of *Geographia* appeared (without maps) in 1475, just two decades after the publication of Gutenberg's Bible, and the same year that presses were being set up for the first time in places like Holland and England. Numerous illustrated editions soon followed, making *Geographia* one of history's earliest bestsellers. The popular 1482 Ulm edition—the one on sale today at Sotheby's—was the first to be printed outside of Italy and helped spur the widespread dissemination of the book, with profound consequences. Wrote the historian Daniel J. Boorstin:

> The revival of Ptolemy . . . would mean the awakening, or the reawakening, of the empirical spirit. Now men

would use their experience to measure the whole earth,
to mark off the known from the unknown, and to desig-
nate newfound places for return. The rediscovery of
Ptolemy was a signal event in the revival of learning
that marked the Renaissance, a prologue to the modern
world.

Spain's King Ferdinand and Queen Isabella ordered their
copy of *Geographia* from a Valencian bookseller. The monarchs'
interest stemmed from their discussions with an ambitious Italian sailor who claimed he could reach the spice-rich Indies by
heading west instead of east. Christopher Columbus had, in
part, based these ideas on what would later prove to be two of
Ptolemy's most famous mistakes: (1) a gross underestimate of
the Earth's circumference, and (2) a gross overestimate of the
eastward reach of Asia. In 1492, after much procrastination and
debate, the Spanish sovereigns commissioned Columbus to sail
with three caravels "toward the regions of India." He never
reached his destination.

. . . $625,000.

$650,000.

$675,000.

$700,000 . . .

The Renaissance cult of acquisition had a dark side as well.
Compulsive collecting could sometimes degenerate into
theft—and some of the great writers and scholars of the age apparently succumbed to this urge. Giovanni Boccaccio, author of
the *Decameron* and a friend of Petrarch, is thought to have pil-

laged a monastic library in his quest to obtain a previously undiscovered piece of classical literature. Poggio Bracciolini, one of the most famous bibliophiles of his time, justified his apparent theft from another monastic library by asserting that the books "were not housed according to their worth, but were lying in a most foul and obscure dungeon . . . a place into which condemned criminals would hardly have been thrust." Needless to say, those in charge of such collections had a different view—as expressed by an inscription at the library of the San Pedro monastery in Barcelona:

> For him that steals, or borrows and returns not, a book
> from its owner, let it change into a serpent in his hand
> and rend him. Let him be struck with palsy, and all his
> members blasted. Let him languish in pain crying aloud
> for mercy, and let there be no surcease to his agony till
> he sing in dissolution. Let bookworms gnaw at his en-
> trails in token of the Worm that dieth not. And when at
> last he goes to his final punishment, let the flames of
> Hell consume him for ever.

. . . $725,000.

$750,000.

$775,000.

$800,000 . . .

The month before the Sotheby's sale, another version of *Geographia* was to have gone on the auction block at Christie's in London. This was an even rarer edition, published in Bologna in 1477 and considered the first-ever printed atlas. Only a few copies survive, meaning the auction would have been an extraordinary event—if it had happened. But it did not happen. It was canceled at the last minute, when Christie's conceded the volume had been stolen nearly a year earlier from France's National Library. That disappearance itself had a bizarre twist: for about three months library officials had simply failed to notice that one of the most important volumes in history was gone. Yet after the theft was discovered in November 1997, it had been widely publicized. It is hard to imagine that Christie's officials did not know about it. Nonetheless, they had apparently accepted at face value the false ownership papers of a Frenchman who brought them the book. According to press reports, the sale had not been canceled until French authorities intervened.

Reading about this debacle a few weeks before the New York auction, I had naturally wondered about the provenance of the Sotheby's *Geographia*. I checked the auction catalog. The most recent owner it listed was a man named Georg Joachim Scherer. He had possessed the book in 1713.

I asked Selby Kiffer of Sotheby's whether his firm would provide me with information about the current owner. No luck. "The confidentiality of both our purchasers and our consignors is something we take seriously," he insisted.

Kiffer, however, was reassuring: "Knowing the consignor of this book as I do, and knowing the history of his family's collecting, there's certainly no doubt in my mind that it's a privately owned copy."

In other words, I would have to take his word for it. I had ab-

solutely no reason to doubt him. But I would have had no reason to doubt Christie's, either.

. . . $825,000

$850,000.

$875,000.

$900,000 . . .

The auctioneer was speaking slowly now, leaving dramatic
pauses to underscore the immensity of the bids. His voice was
calibrated and quiet; it was the only sound in the room. The
competition had come down to two phone bidders, and all eyes
were on their respective representatives, seated on each side of
the podium.

. . . $925,000.

$950,000.

$975,000.

$1,000,000.

$1,050,000.

$1,100,000.

$1,150,000.

The auctioneer waited, but no new offer came. At last he
slammed down his hammer. *Sold.* There was a pause, and then
those in the gallery began to applaud, slowly at first and then
with real enthusiasm. They had just witnessed history. Once
mandatory fees were added, the sale would come to
$1,267,500—a world record for an atlas printed on paper. (In
1990, a scarce copy of the 1482 Ptolemy—printed on an animal-
skin parchment known as vellum—was sold for $1,925,000.)

If Arader was impressed, he didn't show it. I asked him if he was surprised by the sale. "Not a bit," he shot back cheerily. "I *knew* it was going to go for a million. It was a once-in-a-lifetime opportunity for somebody."

Others, however, were clearly in a state of shock. One dealer was slouching out of the gallery when his eyes caught Arader's.

"Twilight Zone," he mumbled, in a dazed, singsongy voice. And then he left, shaking his head.

BY THE END OF THE AFTERNOON, THE BILL FOR Graham Arader's shopping spree at Sotheby's had come to nearly $800,000; roughly half of that was spent to increase his own inventory and half to purchase works on behalf of clients. One of his last purchases of the day was Christopher Saxton's 1579 book, *An Atlas of England and Wales*. It struck me as a perfect symbol of how far Arader had come. More than twenty-five years earlier a "wildly excited" teenager had got his start in collecting by purchasing a map of Dorset from the same edition of Saxton. He had paid $250 for it. Now that same map would sell for $5,500. Back then the entire atlas would have been worth $47,000. Today at Sotheby's, Arader paid $150,000 for it—and, in fairly short order, was asking $450,000.

Looking back on Arader's career, Hugh Kennedy once compared his former boss to Microsoft founder Bill Gates. Each man, he observed, had transformed his respective industry through his own vision and force of will. I thought the comparison was apt in another way, too. Like Arader, Gates had become a lightning rod for resentment. Some was well-justified and well-reasoned—but much was an unfocused rage, expressed in everything from the "Punch Bill Gates" Web page or "I Hate Bill Gates.com" to a real-life 1998 attack on the computer mogul by a group of cream-pie-wielding Belgian anarchists.

It occurred to me that beneath this anger was a lot of envy. We dislike successful entrepreneurs like Arader and Gates, in part, because somewhere in our dark hearts we want to be like them. We wish we had their foresight and confidence, maybe even their ruthlessness, their ability to spot an opportunity and grab it. It angers us that we came on the scene too late, that the doors are now closed, that the money has been made, the power taken, the fame handed out. In many ways these men are heirs to the explorers who first came to this country, brash mercenaries like de Soto and Coronado. We may revile them for their crassness, their avarice, their destructiveness, but who doesn't yearn to have been in their shoes as they strode deep into blank spots on the map?

Our resentment can take warped forms. According to the FBI, computer financial crimes and infrastructure attacks more than quadrupled between 1996 and 1998, as frustrated Bill Gates wanna-bes poured their energies into the electronic underworld of hacking. And Graham Arader wanna-bes? In the words of the map impresario himself: "Someone who comes along who's not well-educated, whose father beat them growing up, who's twisted, but has the same passion for maps I had without any of my advantages, is going to steal them."

DETAILS FROM A 1511
WORLD MAP BY
BERNARDO SYLVANUS.

CHAPTER 4

An Approaching Storm

ZEPHYRUS, FROM THE WEST; BOREAS, FROM the north; Notus, from the south; Eurus, from the east—in Greek mythology, these four deities shared the empire of the winds. Sons of Eos and Astraeus—the dawn and the starry sky—they provided the ancient world with a sense of order. True, Boreas could be nasty and vindictive, chilling the Earth with icy blasts from the north. But on the whole he and his brothers were a hugely positive force, helping to organize human labor, fertilize crops, and orient the routes of navigation. Such powerful icons were these wind gods, in fact, that they reappeared in the Middle Ages and Renaissance, this time on the edges of European maps.

On some maps they took the form of stern old fellows with shaggy beards; on others, flaxen-haired cherubs with playfully pursed lips; on still others, robust men with feathers blooming from their necks, symbolizing their role as God's "wings." Eventually, the original four wind blowers were joined by more of their kind, some shifting positions, others changing names, until there were as many as thirty-two faces surrounding some maps. They were not there merely for decoration, however. Their purpose was all-important: to serve as direction markers. "The Spanish sailors on Columbus' crew thought of direction not as degrees of compass bearings but as *los vientos,* the winds," wrote Daniel J. Boorstin in *The Discoverers.* "Portuguese sailors continued to call their compass card a *rosa dos ventos,* a wind rose. When the religious brotherhood of pilots commissioned the Madonna for their chapel in Cordova, it was no accident that she was *Nuestra Señora del Buen Aire,* 'Our Lady of the Fair Wind.' "

Not all winds were fair, however. Consider the Harpies. They, too, were classical wind deities, but you won't find them on old maps. That's because the sailors and adventurers wouldn't have wanted to think about these tempest-goddesses, much less look at them. The Harpies had the faces of old women, the ears of bears, and the bodies of birds, with long, hooked claws. Their stench was said to be so vile that it would sicken all living creatures. Even their names—Okypete (Rapid), Celaeno (Blackness), and Aello (Storm)—inspired fear. Combining the primitive concepts of wind spirits and predatory ghosts with the actual characteristics of carrion birds, the Harpies brought nothing but disaster and misfortune.

Fair winds or foul? The answer to this question has determined the fate of many a journey—even, perhaps, Gilbert

Bland's own odd version of an epic adventure. And up until that day in downtown Baltimore, we can only conclude that the gods had been extremely kind to our mysterious hero. Zephyrus and his brothers had ushered him safely to and fro across the continent, with stops in Seattle and Charlottesville and Chicago and Vancouver and several other apparent ports of plunder. He had been lucky. Any of a million little things could have failed him—a fumbled razor blade, an inopportune rustling of paper, an unseen security camera—but, miraculously, everything had gone off without a hitch. Or maybe he considered it a matter not of miracles but of his own skill. Maybe, as he emerged undetected from the tenth, eleventh, twelfth, thirteenth, fourteenth, or fifteenth library, he began to feel invincible—or, even more intoxicating to him, invisible. Yes, that seems right: he was suffering from hubris, that prideful arrogance which spelled doom for so many a hero of Greek tragedy, causing him to ignore warnings that might have averted disaster.

Now it was too late for warnings. The winds had changed, and three foreboding figures suddenly stood before Gilbert Bland on the steps of the Walters Art Gallery. To you and me those three would have looked like slightly exhausted security guards. But Bland could be excused if, in his panic and confusion, they suddenly seemed like creatures surreal and terrifying, with whetted talons ready to tear him asunder. When the Baltimore city police arrived at the Peabody Library a short while later, they found a suspect who was sheepish and scared and apparently ready to cooperate. He admitted his name was not

James Perry, as indicated on the University of Florida ID card he had presented to library officials earlier in the day. (Like the name, the card would prove to be a complete fake.) His real identity, he confessed, was Gilbert Joseph Bland, Jr., and he showed police and security officials a Florida driver's license bearing that name. According to library staff members, the man offered no other excuse for taking the maps than that he "just wanted them."

It must have seemed to him just then that there was no way out. He had been caught red-handed with stolen maps, sliced from one of the books he was known to have examined earlier in the day, a 1763 work entitled *The General History of the Late War*. A reliable witness had seen the crime take place, and fingerprint and handwriting evidence would almost certainly further tie him to the scene. The winds of doom now seemed to be blasting through that library, swirling and shrieking. And then, just as it seemed they would lift him off his feet and batter him away to some horrible fate, they suddenly ceased their rage. A calm returned to the room. The police were talking about letting him go.

In his desperation Bland had offered to pay the library—on the spot and in cash—to repair the damaged books, and the cops seemed to think it was a pretty good deal. They had more important things to worry about than Gilbert Bland. Murders had risen nearly 9 percent in the city and its surrounding counties in the first nine months of 1995, compared with the same period of the previous year. Robbery, aggravated assault, and other crime categories were also on the rise. Nor was the

Mount Vernon neighborhood, which had once been among the city's most exclusive enclaves, immune from the problem. Tourists, students, and suburbanites—drawn to the area by attractions such as the famous Peabody Conservatory of Music— were often targeted for robbery or worse.

No wonder the officers did not seem particularly concerned about the meek and skittish man they found at the library. Well-dressed, polite, and obviously humiliated, he looked about as much like a menace to society as the Peabody Library looked like a crack house. And after all, what had he allegedly done? Taken a few pages out of a book? Stolen *four sheets of paper*? There were dangerous people out there—crazy, desperate, dangerous people with guns. This poor guy hardly seemed worth the bother.

But if the police had their own reasons for letting the matter slide, they also had some very practical concerns about how the case would play out, especially given that the suspect lived out of state and might skip out on his bail, with little practical possibility of extradition. "We were advised by the city officers who came that, with this kind of crime, we were running the risk that, if we placed charges, he would make bail and never show up [for trial] and we would get no recompense of any kind," Dennis O'Shea, a Johns Hopkins spokesman, later explained.

Wouldn't accepting the money in lieu of Bland's immediate arrest be easier for everyone involved? The library would get its book repaired; the crook would learn a frightening and costly lesson; the police would have that much less paperwork, that much more time to focus on real criminals. For their part, library officials thought the idea was at least worth considering. They telephoned a lawyer and mulled over their options, Gilbert Bland's fate still blowing in the wind.

HAT TO DO ABOUT BLAND WAS JUST THE LATEST IN A series of difficult decisions that those who ran the Peabody had been forced to make in recent years. For the better part of this century, things at the library had been headed pretty steadily downhill. By 1910 the Peabody Institute was devoting more and more funds to its famous music conservatory, fewer and fewer to the collection. By 1940 new book buying had petered out almost entirely. By 1960 the library—full of dusty old books that no one could check out—had fallen into an alarming state of neglect. "Basically, nobody was using it," explained Cynthia Requardt, curator of special collections at Johns Hopkins University's Milton S. Eisenhower Library. Worse yet, the Peabody Institute no longer had the funds to adequately maintain the beautiful old building.

Cutting its losses, the Institute's board of trustees voted in 1966 to hand the Peabody over to Baltimore's public library system. It turned out to be an ill-suited marriage, one that the library barely survived intact. Shortly after the merger city officials gave serious thought to auctioning off the library's most valuable books, shuffling the rest into the public system, and converting the building into a study hall for high school students. If this caused a furious commotion inside the late Mr. Peabody's grave, it also triggered an uproar in the local academic community, and the scheme was duly abandoned. In the 1970s a fund-raising effort enabled the library to patch the building up a bit, install air-conditioning, and rebind a few hundred books. But as the city's budget tightened bureaucrats decided they could no longer justify the expense of the facility. In 1982 they turned it over to the special collections division at Johns Hopkins University.

This proved to be a much better match in terms of overall philosophy—but the library's financial troubles remained. The university reported that it was pumping between three and four hundred thousand dollars into the facility each year—and even that was not enough. Among other problems, the roof, with its hard-to-repair skylight, was leaking. "The [Grand Stack Room] was opened in 1878 and I'm not sure if the roof had ever been replaced," said Requardt, the Hopkins official who now oversees the Peabody.

To librarians a leaky roof is no small problem. Moisture is the enemy; it causes dampstaining, foxing, and mildew, all of which can stain and destroy paper and some of which are contagious, infecting nearby books with their harmful spores. It takes a lot of effort to defend against these foes, and a lot of money. The temperature and humidity must be precisely controlled—not an easy thing to do in balmy Baltimore, not with a leaky roof, and especially not in a place like the Peabody Library. "It's just an incredibly expensive building to run," explained Requardt. "It's a nineteenth-century building that's on the National Register of Historic Places, so there are certain restraints on what you can and cannot do."

By 1989 the building and the books were at such risk that university officials decided something drastic had to be done. They opted for an approach so controversial it is known in some quarters simply as the "D-word." That stands for deaccessioning, and if you are baffled by the euphemism, think uncollecting or anti-obtaining or conserving in reverse. Put simply, Hopkins was planning to sell rare books in order to save rare books. University officials announced that Sotheby's in New York would be auctioning off ten of the Peabody's books, consisting of sixty-seven volumes, as part of an effort to raise at least $2 million toward a $4 million endowment fund for the library. Included in that group were two copies of Hartmann Schedel's 1493 *Nuremberg Chronicle,* which contains one of the most important early printed maps of the world, as well as works by the photographer Edward Curtis and the nature artists John James Audubon, Mark Catesby, and John Gould.

Hopkins officials insisted they had no choice: the very future of the Peabody Library was at stake. "Our goal is to try to find ways to preserve [the library]," the university provost, John Lombardi, told the Baltimore *Sun*. He claimed the sale would maintain the library "with a minimum amount of change and a maximum amount of preservation and access." Officials also were quick to point out that all the books to be sold were duplicated in other Hopkins collections. Nonetheless, the announcement touched off a firestorm of criticism, typified by the blunt words of John Burgan, chief librarian at the public library in Hartford, Connecticut: "It borders on the rape of a great library collection."

To Burgan and other foes of the move, deaccessioning in order to preserve the library was like burning clapboards in order to warm a house. "The Peabody Library is one of the great treasures of the United States, and once you sell off the rare books you might as well close the library, because that's the uniqueness of it," declared Regina Soria, professor emeritus of modern languages at the College of Notre Dame.

But deaccessioning accounted for only part of the furor. Other museums and libraries had sold off some of their collections to make ends meet; in many people's minds the practice was ugly and unfortunate, but perhaps not unforgivable. No, what critics really found offensive was that Hopkins planned not just to sell historic old volumes but to cut them up beforehand. That's right: book breaking—not by some profit-obsessed dealer but by one of the oldest and most respected libraries in the United States. One of the books scheduled to go under the knife was Audubon's *Birds of America*. The Peabody's copy was extremely rare—one of only two in the world that contained Audubon's signature on all four volumes. Nonetheless, Hopkins and Sotheby's planned to gut the four-volume set so that its 435 prints could be sold separately. "That's disgraceful," Arthur Gutman, a member of the Johns Hopkins University Library Advisory Council, told the *Sun*. "It ought to be sold, if it has to be sold, to a sister institution, and not broken up."

Despite all objections, the sale went forward, raising a total of $2.4 million and allowing Hopkins to keep the building and the collection open to the public. One print alone, "American Flamingo" from *Birds of America*, fetched $66,550—a record at the time for a single Audubon. The buyer was a fellow named Graham Arader.

But while such head-snapping sums were a godsend for the Peabody in the short run—enabling Hopkins to pay for everything from staff salaries to upkeep of the physical plant—they posed a serious long-term threat to libraries across the country, which faced increasingly tough security concerns as the value of their collections skyrocketed. The Peabody itself had been saved, but its once stellar reputation had been noticeably damaged.

Now, as they questioned Gilbert Bland, Peabody officials faced yet another tough choice. Had they been aware that their decision would touch off another controversy, they would undoubtedly have given it more thought. But they had no notion of what—and whom—they were dealing with. "Frankly, my sense of it was that we had caught a bumbling, stumbling amateur," said Frederick DeKuyper, associate general counsel for Johns Hopkins University, whom Peabody officials consulted in the immediate aftermath of Bland's detention. "None of us at the time knew of his past activities."

After conferring with security officials, DeKuyper signed off on a plan to forgo Bland's arrest in return for seven hundred dollars in damages. The suspect, who was reportedly carrying large amounts of cash, was more than happy with the deal. With the gods apparently smiling down upon him once again, he fled the library in a hurry—too much of a hurry, as it turned out.

*H*ERE'S THE THING ABOUT THE HARPIES: THEY SCREW with your head. Known by ancients as the Snatchers because they loved to steal children, souls, and other precious things, they were brilliant at torturing their victims. When Zeus condemned the soothsayer Phineus to everlasting hunger, it was the Harpies who served as enforcers. Each time Phineus attempted to eat, the Harpies would swoop in, stealing his food or fouling it with excrement just as he was about to put it in his mouth.

And now—just as Gilbert Bland was so close to freedom that he, too, could practically taste it—a similar fate befell our hero. Was it the Harpies who placed that notebook out of his mind and out of his reach as he left the Peabody Library? If so, it would be perfectly in keeping with their traditional role as tormentors of those with wicked obsessions. No matter: whether by intervention of the gods or simple forgetfulness, the fateful fact remains that Bland left his notebook behind. Within minutes of the thief's departure, the Peabody's security chief, Donald Pfouts, noticed the book and decided to give it a closer examination. He quickly made a discovery that would send shock waves through libraries all over America.

To ILLUSTRATE THE SECURITY CHALLENGES FACED BY IN-stitutions such as the Peabody Library, let us briefly return to Baltimore's Washington Monument, that "towering main-mast," which Bland had passed under during his unsuccessful escape attempt. Completed in 1829, the monument is the na-tion's oldest public memorial to George Washington, and it holds a central place in the city's history. In the words of one tourist brochure, "It put Baltimore on the world map." Experts still consider the design by the architect Robert Mills to be one of the nation's foremost examples of Romantic neoclassi-cism—but average people are passionate about it, too. When it needed repairs in the 1980s, and the cash-strapped city ran out of money to do the job right, a group of business leaders stepped in, declaring the monument a "national treasure."

Now let us consider a different national treasure, this one also the work of Robert Mills. I am not referring to his *other* Washington Monument, the famous one in the nation's capital, or to his Treasury Building, the famous one on the back of a ten-dollar bill. I am referring to his *Atlas of the State of South Carolina*. It turns out that, in addition to being a noted architect, Mills was an important mapmaker. His 1825 masterpiece was not only a beautiful piece of art in its own right but a work of real historical importance—the first state atlas ever produced in the United States. Yet all national treasures are not the same. We can be sure that if the monument were defaced, there would be a public outcry. But few people would complain if someone desecrated a copy of the *Atlas of the State of South Carolina*. It's a pretty fair bet, in fact, that no one would even notice the crime had taken place.

But if the public at large knows virtually nothing—and cares even less—about the *Atlas of the State of South Carolina,* a small number of map collectors and dealers value it a great deal. Today a copy of the atlas in excellent condition might sell for upwards of $30,000. A single map of Charleston County from the atlas might fetch $2,000 or more. And, unfortunately, what collectors and dealers are willing to pay for, thieves are willing to steal.

So it was that, as he flipped through Gilbert Bland's notebook, Donald Pfouts came across the following passage:

For MD Dealer
1. *Currier & Ives (90%)*
2. *Kellogg*
3. *Haskell & Allen*
4. *Baillie*
Mills County Atlas of <u>*S.C.*</u>

Pfouts could not help but recognize the disturbing implications of such words. Almost every page was filled with lists, most of which contained the names of cartographers and the titles of specific maps (and a few of which, such as the preceding example, also included the names of artists and their prints). Next to many of these entries were prices. Pfouts quickly came to a startling conclusion: the notebook was essentially a hit list. Worse yet, there was ample evidence that, far from being an isolated incident, as library officials had assumed, Bland's visit to the Peabody was part of something much bigger. The book contained the names—and, in some cases, addresses—of other libraries where specific maps could be found. Moreover, folded into its pages were informational materials from institutions such as the University of Virginia's Alderman Library, a disturbing clue that other thefts may have already taken place. Then there were those ominous words "For MD Dealer"—a possible implication that the manuscripts were stolen on commission. And, finally, there were a number of potentially incriminating notes, which Bland had apparently written to himself or perhaps passed to an unknown accomplice. They strongly hinted that the intruder was a man on a mission, operating in a highly organized and deliberate fashion. "Can't these people leave? I can't do it now. OK now," read one of the neatly written passages. "These 2 are done now. Thank God!" read another. "Yes she's really slowing me down! Fat Bitch," read a third. And a final one: "The Bowen Atlas—of all the bad luck—What is going on here. Am I not going to get these Bowens? What [will] become of me?"

It was a question that was beginning to interest Peabody librarians a great deal—especially after they went back through their own records and discovered that more maps were missing from other texts that Bland had allegedly handled, during both this visit and one the previous September. Did he get his

Bowens? Apparently so. Four plates were missing from the now decidedly incomplete *Complete Atlas or Distinct View of the Known World* by the eighteenth-century cartographer Emanuel Bowen, along with material from works by Mathew Carey, Jacques Nicolas Bellin, Entick, and Pierre-François-Xavier de Charlevoix. In all, twenty-seven plates were missing from the Peabody alone. Had other institutions suffered similar losses? It was now becoming frighteningly clear to Hopkins officials that they might be dealing with a crime spree. Cynthia Requardt—who had not been in on the decision to free Bland—quickly began calling the libraries that appeared to have been targeted in the notebook. Then she went to her computer and sent out a message over ExLibris, an electronic discussion group for those interested in rare books and special collections:

On December 7, Gilbert Joseph Bland, Jr., was apprehended removing maps from eighteenth century books at the George Peabody Library of The Johns Hopkins University in Baltimore.

Bland was using the alias James Perry. He is a white male, 46 years of age, 5′ 9 or 10″, with light brown hair (receding) and a light brown moustache. A photograph is available.

When apprehended, Bland presented a Florida driver's license. In lieu of pressing charges, the library accepted payment for damages, and Bland was released. Since his release, we have reason to believe that Bland has visited other research libraries in the mid-Atlantic region.

For Requardt, Pfouts, and other Hopkins officials, there was little to do but wait—and hope that their worst fears would not be realized.

CHAPTER 5

How to Make a Map,
How to Take a Map

*I*T TAKES THOUSANDS OF YEARS. A SINGLE map, observed Denis Wood in *The Power of Maps,* is not a self-contained document but a compilation of what *"others have seen or found out or discovered, others often living but more often dead, the things they learned piled up in layer on top of layer so that to study even the simplest-looking image is to peer back through ages of cultural acquisition."*

Geographic discoveries are part of it, of course, but before you can explore the land you have to explore the heavens. "Progress in the science of cartography has never moved ahead of developments in astronomy," wrote the map historian Lloyd A. Brown, "and our world map of today has been made possible largely because of the high degree of accuracy achieved by

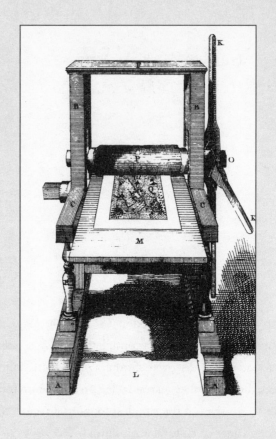

AN EIGHTEENTH-CENTURY
PRINTING PRESS.

astronomical observers." Coming to grips with this notion takes millennia all by itself, and then you still need the right tools. The sundial in ancient Babylonia and Egypt, the magnetic needle in ancient China, the astrolabe in ancient Greece, the cross-staff in the fourteenth century, the backstaff, telescope, and theodolite in the sixteenth century, the quadrant in the seventeenth century, the octant, sextant, and chronometer in the eighteenth century—these instruments make measuring the size and shape of the world possible. Nonetheless, they can only provide bits and pieces of information at any one time. In drafting your map, you will also need to depend on eyewitness reports of sailors, soldiers, merchants, adventurers, braggarts, and scam artists. You will have to consult the work of other cartographers and, when all else fails, rely on educated guesses and the occasional time-honored assumption (read: myth).

It's an incredibly difficult job. Let's say you're working in the seventeenth or eighteenth century, an age in which mapmakers were sometimes referred to as "world describers." In geometry, *describe* means to draw or trace the outline of something; in poetry, it means to get at the essence of something, to bring it to life in a way that's both startling and beautiful. You've got to do both kinds of description—and do it in a medium that's partially visual, partially mathematical, partially textual, a complicated miscellany of scale, orientation, projection, grids, signs, symbols, lines, colors, words. It's often been said that mapmakers combine science and art, but there's far more to it than that. Joan Blaeu, a legendary Dutch cartographer of the seventeenth century, wrote that "maps enable us to contemplate at home and right before our eyes things that are far away." Making the distant immediate, the unseen visible—that's far beyond science or art. That's alchemy.

It may take you months, even years, to draft a single map. It's not just the continents, oceans, mountains, lakes, rivers, and political borders you have to worry about. There's also the cartouche (a decorative box containing printed information, such as the title and the cartographer's name) and an array of other adornments—distance scales, compass roses, wind-heads, ships, sea monsters, important personages, characters from the Scriptures, quaint natives, menacing cannibal natives, sexy topless natives, planets, wonders of the ancient world, flora, fauna, rainbows, whirlpools, sphinxes, sirens, cherubs, heraldic emblems, strapwork, rollwork, and/or clusters of fruit.

And once you're finally finished describing the world, you have to transfer it to a copper plate for printing. This, too, takes time. You select a piece of copper that's not too soft, not too hard, and you planish it with a hammer to smooth it out and to ensure that it is firm and free from holes or flaws. Then you have to polish the smoothest side to a mirror-smooth finish, first with a piece of grinding stone and water, next with a pumice stone, then with a hone and water, then with hardened charcoal, and finally with a steel burnisher. Perfection is a must: any minor scratch might show up as a line on your map, possibly causing the accidental subdivision of one small nation into two smaller nations.

Next, you transfer the design of your map to the plate. This is a difficult process that involves heating the plate, spreading a layer of wax over it with a feather, then laboriously tracing the map, in reverse, onto the wax coating. Once that's done, you're ready to cut the design into the plate. You might do this by etching—using a needle to scratch the outlines of the map through the wax coating, then pouring acid into the needle marks to burn your design into the plate. More likely, however, you would engrave the plate—cut the design into it by hand—using a variety of tools: the burin (also known as the graver), the tint tool, the scauper, the threading tool, and the roulette. This is precision work, and it usually requires a number of people and a great deal of time, not the least because at several intervals in the process it is proofed and then corrected, by meticulously smoothing over the mistakes with a burnisher and then reengraving.

Next comes the ink. You spread it over the plate, then carefully wipe off the excess until the only remaining ink sits in the engraved grooves. You heat the plate until it is warm, lay it out on the printing press, and place a sheet of thick dampened paper over it. When the press is tightened, the moist paper draws the ink from the incised lines of the plate—and, at last, you are looking at a printed map.

But you might not be done. You might also need to color your engraved map—and there is no machine to help you with this. You must do it by hand, brushing watercolors onto the freshly printed map—meticulous detail work that, if done wrong, ruins all the labor that came before it, and, if done right, gives birth to bright-hued oceans and vibrant nations. On a particularly sumptuous map, you might also need to add gold-leaf highlights.

All that, and you've made only one map. What if you're making a whole atlas? The scholar C. Koeman once tried to estimate how long it took for the great Dutch mapmaker Joan Blaeu to print his famous *Atlas Major* of 1663. Assuming a relatively small press run of three hundred copies for each of the first three editions of the atlas (Latin, French, and Dutch), Koeman concluded that the composition (or typesetting), with eight full-time employees, would take 1,000 working days; the letterpress printing (of the text parts of the book), involving nine printing presses, would take 330 working days; the copperplate printing (of maps and other graphic elements), involving six printing presses, would take 900 working days; and the binding, involving three employees, would take 300 working days. "The planning involved in printing the three editions . . . within a span of three or four years," Koeman concluded, "exceeds the range of our imagination."

Making a map, in short, is a painstaking process, requiring extraordinary skill on the part of everyone involved, but there is a huge demand for your work. Members of a later generation will smugly call their era the Information Age, but it will be nothing compared with this one. The printing press is changing many aspects of life, none more so than the way people envision the Earth. This breakthrough invention not only allows for maps to be produced in accurate and standardized ways but makes possible the widespread dissemination of geographic images. For the first time the whole world is able to see the world as a whole.

PART TWO

*I*T TAKES NO TIME AT ALL. AND YOU NEED only one tool. An out-of-the-way spot is preferable but not necessary: with enough skill and daring, you can do this right in the middle of a busy rare books room. If making maps requires real magic, taking them involves only sleight of hand. Just listen to how Gilbert Bland did it. "When no one was looking he would proceed to take out a single-edged razor blade, like you would use to scrape stuff off glass," explained Lieutenant Detective Clay Williams of the University of North Carolina Department of Public Safety, who talked to Bland about the heists. "He could put the razor under his fingers so that you never really saw it. You just saw him take his hand and go down from the top of the page to the bottom. It would appear to be nothing unusual—maybe like he was just scanning text. But he would actually be cutting out the page. The whole operation would take just a matter of seconds."

Slide the page into your coat and you're through the door, hundreds of years of history gone in less time than it took the ink on the map to dry.

*T*HE HARD PART ISN'T THE ACTUAL STEALING, IT'S THE getting in and getting out. Even at the least secure of libraries, you must present your identification, sign a register, perhaps even pass by a security guard and answer a few questions about your reasons for being there. This is usually when you feel the most trepidation and anxiety, like a smuggler at a border crossing. In fact, you are crossing a kind of border. As a map aficionado, you know about borders. You understand that they are not just lines on a piece of paper but powerful psychological metaphors. "The crossing of the border symbolizes transgressing against moral commands or trespassing into forbidden territory," observed Avner Falk, a clinical psychologist interested in the relationship between maps and the mind.

And make no mistake: a rare books room *is* forbidden territory. Scholars believe that libraries evolved from ancient "temple collections" that housed a religion's most important texts. "The theological collection was kept in a sacred place, and presided over by a priest," wrote Michael H. Harris in *History of Libraries in the Western World*. "Only the most important of the temple officials might have access to this library." In Judeo-Christian tradition, the temple collection was exemplified by the Holy of Holies, the innermost sanctum of the Jewish tabernacle and home to the Ark of the Covenant, which contained the original stone tablets of the Ten Commandments. The Ark was considered so sacred that God would smite dead those who touched it.

You have come to desecrate the modern equivalent of this temple, a largely off-limits chamber where our culture stores its rarest and most valued documents. But while penetrating this forbidden place no doubt fills you with apprehension, it might also give you a huge thrill. "Some people seem to seek the exhilaration of border crossing, and when conventional borders no longer seem exciting, more imposing ones are sought. . . . To even possess the ability to cross a dangerous border may be a necessary potency for some individuals," wrote the psychiatrist G. Raymond Babineau.

From 1967 to 1970 Babineau was the chief of psychiatric services at the U.S. Army Hospital in Berlin. There he had the opportunity to observe many "compulsive border crossers"— restless souls who repeatedly made dangerous journeys over the Iron Curtain in order "to be rid of one psychological state and catapulted into a newer and better one." Maybe you're driven by similar impulses. Maybe, like those Babineau interviewed (and, as court records indicate, like Gilbert Bland), you suffer from depression. Perhaps strolling through the doors of the library allows you to leave your feelings of low self-esteem and ineffectuality behind for a few moments and enter a world in which you are skilled and powerful. You may also, like Babineau's border crossers (and like Bland), come from a broken home. If so, your visit to the library might be what Babineau called a "search for identity." History shows that people often take on new personae when they cross borders, from the Old Testament, in which Jacob became Israel after crossing the Jabbok River, to Ellis Island, where countless immigrants likewise were given new names. It may be that by entering the library you, too, are trying to be reborn, just as Gilbert Bland became James Perry when he walked through the door.

In addition to the border crossers that Babineau studied firsthand, he also examined the case of a young man named Lee Harvey Oswald. You may have more in common with him than you think. True, stealing a few maps is a long way from killing a president. But look at *The Warren Commission Report*. See if you don't recognize something of yourself in it:

> Perhaps the most outstanding conclusion of such a study is that Oswald was profoundly alienated from the world in which he lived. His life was characterized by isolation, frustration, and failure. He had very few, if any, close relationships with other people and he appeared to have great difficulty in finding a meaningful place in the world. . . . [His wife] Marina Oswald thought that he would not be happy anywhere, "Only on the moon, perhaps."

Maybe you steal maps because you're searching for a home. Or perhaps you simply crave the dark joy of appropriation. Because that's another thing about borders: they are mighty tools of theft. "Countless governments have used map boundaries to subdue, displace, or annihilate native peoples," noted the geographic scholar Mark Monmonier in *Drawing the Line:*

Tales of Maps and Cartocontroversy. Talk about magic: with one teetery line, what was *ours* suddenly becomes *yours.* You understand how easy this is. That's why you are adding a new border to a map at this very moment. But you are not holding a pen. Your border is limned with cold steel.

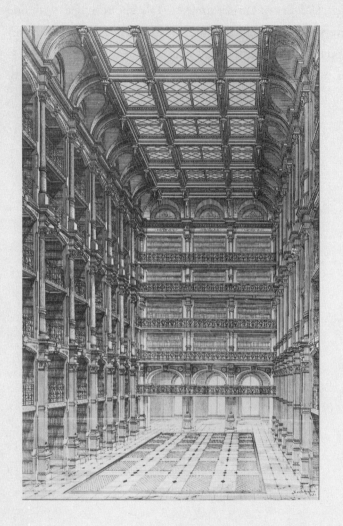

THE PEABODY LIBRARY'S GRAND STACK
ROOM, FROM AN 1879 ENGRAVING.

CHAPTER 6

The Invisible Crime Spree

*T*HE GHOST OF LLOYD A. BROWN WAS NOT pleased. He floated invisible amid the mote-speckled air of the Grand Stack Room, nervously chewing his pipe or rubbing his bald pate, his bushy eyebrows arched with tension. In life, Brown had been head librarian at the Peabody. While there he introduced the first modern ventilation system to the building, overhauled the outdated card catalog, launched a campaign to degrime hundreds of books, and, as part of that effort, took it upon himself to personally inspect the contents of every shelf on all six tiers. But his biggest accomplishment, during a tenure that ran from 1942 until 1956, was his efforts not as a librarian but as an author. His groundbreaking book, *The Story of Maps,* published in 1949, set the standard for all cartographic histories to come—and is still in print more than fifty years later. With no formal training in cartography, Brown had not been an obvious candidate to write such a volume, but he'd had a couple of advantages going for him. The first was an obsession with old maps. "If you get bitten by a flea, I guess you have to live with it," he once joked. The second was his own library's extraordinary collection. In researching *The Story of Maps*, Brown had needed to consult more than five hundred books. He delighted in recounting that all but ten of them were found on the shelves of the Peabody.

Since his death in 1966, Lloyd A. Brown had led a happy spectral existence amid his beloved books. Or at least that's the way I imagine him. What a wonderful thing reading must have been, when unencumbered by the earthly pressures of time! One could linger on every word. No volume was too long, no passage too dense, no subject too unfamiliar, no metaphor too obscure, no foreign script too unintelligible. One need not despair, as in life, about the books one would never have the opportunity to open or reopen. Lloyd Brown had gone to heaven, and it was called the Grand Stack Room.

And so things might have remained, if not for the intruder— the hated one who crept into the library one day, seated himself at one of Brown's favorite old tables, and, as the ghost hovered helplessly above, began to slice up books. Worse: map books— the ones Brown had cradled so often, so tenderly, for so many years, the ones that had succored his intellect and imagination, the ones that had given him a sense of purpose, the ones that had ensured his name would live on long after he passed into the realm of the spirits. It was as if that razor blade was not just cutting paper but severing Brown's ties to the realm of the living. He raved, howled, shook incorporeal fists, shed invisible tears. None of it helped. I imagine that even after the man was apprehended, Lloyd A. Brown's ghost continued to wander disconsolately, muttering curses heard by no one—a spirit no longer at rest. The intruder had caused him to understand that while he would haunt this piece of air forever, all he had worked so hard to create could disappear, page by page, book by book, shelf by shelf, wall by wall, until the Peabody was nothing but a landmark on some old city plan, the last faded trace of his life. Of what use is eternity without the past?

MONG THE LIVING AT THE PEABODY, THE MOOD WAS not much brighter. In the days after Cynthia Requardt's Internet posting about Gilbert Bland, all news was bad news. Requardt had sent the message as "an attempt to warn people to watch out for Bland in the future," she later recalled. In so doing, however, she had unintentionally opened Hopkins officials to criticism—a torrent, as it would turn out. One writer after another filed messages to the ExLibris news group, lambasting the university for allowing Bland to go free. Such criticism pained Requardt. She would later concede that she was simply "not prepared" for the "candid assessments" of her peers. Yet as much as the attacks stung, she was even more devastated by the reports arriving from other libraries. James Perry had been to the University of Virginia. James Perry had been to Duke University. James Perry had been to the University of North Carolina and to Brown. At all of those stops, books handled by the mystery man now appeared to be missing maps and prints.

Perry's visit to the Regenstein Library at the University of Chicago on October 31, 1995—Halloween—came five weeks before Bland was detained in Baltimore. Walking calmly into the library, he sat down in the special collections room and opened one of the Western world's most extraordinary texts: a 1584 edition of *Theatrum Orbis Terrarum,* "Theater of the World," compiled and edited by Abraham Ortelius, the father of modern geography. Ortelius, a Flemish cartographer, lived during an unprecedented period of discovery. Columbus had landed in the Americas, Magellan's expedition had circumnavigated the globe, Copernicus had made his case for a sun-centered universe. Yet cartography was behind the times. Maps came in a slapdash variety of sizes and styles, many of them based on the ideas of Ptolemy and other ancient geographers, who, of course, had known nothing about the existence of North or South America. Ortelius set out to change that, painstakingly collecting the finest maps of places throughout the known world, then bringing them together in a uniform size and format. Originally published in 1570, *Theatrum Orbis Terrarum* was the first modern atlas. Ortelius put the whole world at the fingertips of the traveler in a standardized fashion—a milestone in the history of human imagination.

Yet now the great master's text had wound up in the hands of a kind of Anti-Ortelius, a professional scatterer of maps and destroyer of books. Bland apparently paged through the volume until he came to a map labeled *La Florida,* the first widely available map of the broad region that is now the southeastern United States. Ortelius added it to *Theatrum Orbis Terrarum* for the first time in this 1584 edition. And there it remained for the next 411 years—until the intruder allegedly flashed out his razor blade.

Although the book measures seventeen inches by twelve inches and its pages are so thick that they literally rumble when turned, Bland is believed to have removed *La Florida* and two more plates from that same atlas, as well as ten maps from another book. The Regenstein's special collections room is a kind of fish tank built expressly for security: its walls are made of

"LA FLORIDA" MAP FROM THE LANDMARK ORTELIUS ATLAS, *THEATRUM ORBIS TERRARUM.*

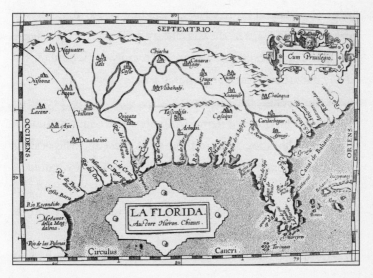

glass, and no briefcases or pens are allowed inside. Yet Bland seems to have sneaked the thirteen plates into his clothes and walked out undetected. For good measure, he also altered a librarian's pencil-written inventory at the front of the Ortelius book, making it appear that the maps he took had been missing for years.

But that wasn't his only alleged theft during his brief Chicago stay. Only the day before he'd paid a visit to Northwestern University's Charles Deering McCormick Library of Special Collections, a churchlike chamber with arched windows and cluttered old bookshelves adorned with busts of Dante, Einstein, Dylan Thomas, Samuel Johnson, and Ezra Pound. The curator R. Russell Maylone remembers the visitor as "the proverbial man in the raincoat" with "a pile of books on the table spread out in a not very orderly fashion." That day Bland is believed to have removed six separate maps from the pages of several antique atlases, including a 1681 map of New York and three maps of the Caribbean. As Perry got up to leave, Maylone said, "I hope you found what you were looking for."

He had, thank you. He certainly had.

As the weeks rolled on, more and more institutions reported visits from the dreaded Perry. Ultimately the count would rise to nineteen libraries, most of them in the East and Midwest but one in Seattle and two in British Columbia. The list would include such institutions as the University of Delaware, the University of Florida, Washington University in St. Louis, and the University of Rochester.

No one had called the police. No one, apparently, had noticed that the maps were gone. It was an invisible crime spree, hidden amid the seldom-opened pages of centuries-old books. And its perpetrator was the invisible criminal. Beyond a general physical description, the people who met him would later describe him in only the vaguest of terms: "clean-cut," "quiet," "polite," "mild-mannered," "nondescript and noncommunicative." Many simply had no memory of him at all.

HAT A VAPID JOB TITLE OUR CULTURE GIVES TO those honorable laborers the ancient Egyptians and Sumerians variously called Learned Men of the Magic Library, Scribes of the Double House of Life, Mistresses of the House of Books, or Ordainers of the Universe. *Librarian*—that mouth-contorting, graceless grind of a word, that dry gulch in the dictionary between *libido* and *licentious*—it practically begs you to envision a stoop-shouldered loser, socks mismatched, eyes locked in a permanent squint from reading too much microfiche. If it were up to me, I would abolish the word entirely and turn back to the lexicological wisdom of the ancients, who saw librarians not as feeble sorters and shelvers but as heroic guardians. In Assyrian, Babylonian, and Egyptian cultures alike, those who toiled at the shelves were often bestowed with a proud, even soldierly, title: Keeper of the Books.

I think most librarians, especially rare books librarians, still secretly view themselves in this way. I met dozens of them as I retraced Bland's footsteps, and in the end I came to the conclusion that most view their work as what the scholar and novelist

Umberto Eco, writing about librarians of an earlier age, called a "war with the forces of oblivion." I think they wage this war because, like the historian Barbara Tuchman, they believe that "books are humanity in print" and that, as the Keepers of the Books, they are safeguarding not just pieces of paper but mortal flesh. The stakes are high: if they fail, the past dies.

Not that the librarians themselves are likely to talk in such terms. I found them to be a guarded, self-effacing, and often sardonic lot, a group decidedly short on Graham Arader types. As a result, I had a tough time getting a handle on what made them tick. I confess that this did not bother me too much in the beginning. Of the many mysteries surrounding the Bland case, the Secret Life of Librarians was not exactly the first I yearned to ferret out. But as the months wore on, and my respect for librarians grew, so did my curiosity about their craft. Yet, try as I might, I was unable to gain a good feel for their esprit de corps until I got to know Gary L. Menges, one of the last librarians I interviewed.

When I first contacted Menges over the phone, he sounded just as you might expect a librarian to sound, from the nasally voice to the precise, even persnickety, way of putting words together. But when I traveled to Seattle for a meeting with him at the University of Washington campus, I was greeted by a man much different than the one I had imagined. Far from being a tepid biblio-bureaucrat, Menges seemed full of vitality and purpose. He later told me that he was sixty-one years old; I had assumed he was a good fifteen years younger. His most memorable attribute, other than a tastefully radiant red-and-orange necktie, was a plush, meticulously maintained white beard. He reminded me of a very trim, very tidy version of Santa Claus, complete with a twinkle in the eye and a disarmingly robust laugh.

Menges led me to his basement office at the Kenneth S. Allen Library, where he worked as the head of the Special Collections and Preservation Division. On one wall hung a series of New Year's cards, which the department sent to its donors and supporters every winter. Each contained the work of a local artist, accompanied by a quotation about books. One read: "The world exists in order to become a book."—Stéphane Mallarmé. Another: "I have always imagined that paradise will be a kind of library."—Jorge Luis Borges. A third: "Handle a book as a bee does a flower, extract its sweetness but do not damage it."—John Muir. Menges told me the idea for the cards had been his; it would soon be clear to me that the sentiments were as well.

I had not come to talk about greeting cards, however. In the middle of the room sat a cart, holding a large, leather-bound volume.

"Ah," I said. "This must be *the book*."

"Yes," he replied. "I thought you'd want to see it."

It was true. I had already heard much about this particular book, and I was anxious to examine it for myself. But I also wanted to observe Menges as *he* looked at it, touched it, talked about it. Menges, I knew, had gone to extraordinary lengths to obtain and preserve this volume. I had come here to get a first-hand feel for one librarian's intense devotion to a single text.

The saga began in 1990, when Menges was contacted by a

University of Washington alumnus from Spokane. The man owned a collection of old books that had once belonged to his grandfather, a history professor at Northwestern University. He hoped the library might have use for the volumes. Menges, of course, welcomed the offer, but he realized, too, that nothing might come of it. Special collections libraries are, by their very nature, highly selective—and often the books people want to donate are either of little historical interest or already on the library's shelves. But you never know when and where you're going to strike gold. Once, for instance, somebody had fished some old architectural drawings out of a dumpster and brought them to the library. They turned out to be the original plans for some important local buildings.

The Keeper of the Books must always keep his mind open. Menges hopped on a plane for Spokane, having concluded the man's collection was at least worth checking out in person. It was there that he first laid eyes on the book. Its ornately decorated leather spine was badly worn, but the gold-leaf title was still legible:

OGILBY'S

AMERICA

Menges did not need to be a rare books librarian to see that it was a beautiful work. That would have been obvious to anyone taking a cursory perusal of its 122 superb engravings, including many gorgeous maps. But Menges also recognized it as a historically important volume. Its author, the Scottish cartographer

John Ogilby, had served as royal cosmographer and geographic printer under England's King Charles II and had published *America* in 1671 as part of a series of works about other lands. It was not a particularly original work (most of the information was, in fact, lifted from other sources) nor a particularly accurate one (placing Montezuma's Aztec empire, for example, in Peru). Nonetheless, it had the distinction of being the first encyclopedia of the New World to be published in English. Moreover, Menges knew that Ogilby had authored several other map books, most notably *Britannia,* the first-ever national road atlas. Originally published in 1675, *Britannia* was later reprinted in a number of editions—including a 1762 version, revised by John Senex and retitled *The Roads Through England Delineated,* which now sat on his library's shelves. If nothing else, Menges relished the opportunity to bring two of Ogilby's most important works together under one roof. "This obviously was one of the nicest books—if not the nicest book—in the [Spokane] collection," he recalled. "I, of course, expressed a strong interest in it."

He understood, however, that in the thorny world of donor relations, interest does not always translate into acquisition. The Keeper of the Books must schmooze. Menges was not surprised that the man did not immediately relinquish the book, wanting some assurances before making the gift. "Because it had belonged to his grandfather, and there was this family tie to it," he explained, "he was concerned that it be in a good home."

Luckily, Menges was in an ideal position to talk about good homes just then. His Special Collections and Preservation Division was setting up shop in the brand-new Kenneth S. Allen

Library. The building had been named after a former university librarian, who (like most former university librarians) would not have had a large edifice built in his honor if not for one fact: he was the father of Microsoft cofounder Paul Allen. Thanks in large part to the computer magnate's $10 million gift (the largest in university history to that time), the architect Edward Larrabee Barnes had been able to fill the new facility with all sorts of state-of-the-art technology—from special filters on lights and windows, designed to guard books against damaging ultraviolet rays, to a computer-monitored, climate-controlled vault where rare works, such as the Ogilby, were to be stored.

"The book's going to be here forever," Menges could confidently promise the alumnus from Spokane. "It's going to be well cared for, and it's going to be used."

In making such assurances, Menges was putting not just the library's reputation but his own on the line. If anything happened to the book, he would be stuck with the unenviable task of explaining things to the donor—not to mention administrators, journalists, and, worst of all, other potential givers, who might consequently hesitate to put their own beloved volumes in his care. But his obligations did not end there. The Keeper of the Books must answer to the past and future, as well as to the present. In assuming responsibility for the book, Menges would be accountable to everyone who had ever read it, studied its maps, placed it lovingly on the shelf. They had kept the volume intact for more than three hundred years, and Gary Menges owed it to them to do no worse. Yet he had an even bigger responsibility—his most solemn duty of all, in fact: to keep the book in good condition for generations to come, readers not yet born.

This burden was one that Menges was only too happy to shoulder. I could see that right away as I sat in his office that morning—and it occurred to me that the donor from Spokane must have seen it, too. As I was, he had probably been reassured by the librarian's unpretentious enthusiasm for the book, all the more convincing for its lack of bluster, and by the careful but confident way Menges held it in his hands. And perhaps he was won over by the way Menges looked at it—not with the covetous stare of a lover, not even with the overly jealous eye of a parent, but with the proud and protective gaze of a good grandparent. The longer I talked to Menges, the more I suspected that the man's final decision to donate the book had less to do with the high-tech frills of the new library than with the quiet charisma of a certain old-fashioned librarian.

After the negotiations were completed, the legalities hashed out, and the paperwork signed, the book arrived in Seattle. But Menges's work was just beginning. His next job was to get the volume physically ready for the shelves—no small task. "This particular copy was not in the best condition," he explained. "The text block [the part made up of the actual pages] was intact, but the bindings had worn out. Most bindings do, if the book is heavily used for three hundred years."

As he told me this, Menges reached into a file and pulled out a couple of photographs, taken of the book when it first came to the library. It looked as if the volume had been chained to the back of a pickup and taken for an off-road adventure. The bind-ing—in layman's terms, the cover—was ripped, stained, and generally worn-out. The leather itself was cracked and brittle. The corners were frayed. The frontboard—the part of the cover you would open in order to read the book—was com-pletely detached. There were other problems as well, including some ripped maps. All told, Menges said, the estimated cost of repairing the damages had come to almost a thousand dollars— no small sum to plunk down on a single book, but manageable if you've got it in your budget. The trouble was, Gary Menges didn't have it in his budget. He would have to raise this money on his own.

One thing Menges did have, however, was a plan. He called it Save a Book. He had launched the program a year earlier, using a simple formula. First, select a beautiful old volume in a frightful state of disrepair. Then, put it in the display case that people nearly stumble over every time they walk into the read-ing room. Next, print up a bunch of flyers extolling the vol-ume's aesthetic and historical import. And finally, beg, cajole, and beg some more.

Menges has a subtle gift for this art—as I learned when, in the midst of our interview, I found myself reaching for my checkbook to help out poor old Sir Walter Raleigh, the Save a Book poster child du jour, whose *History of the World* had fallen on hard times. "All sorts of people donate," Menges cheerfully explained, moments before accepting my modest alms. "On our first Save a Book effort [for Hartmann Schedel's *Nuremberg Chronicle*], the donors ranged from the president of the university to a student who came in and gave me a dollar, saying that was all he could afford. I assured him that his dollar was most welcome."

The Keeper of the Books must enlist others to the cause. But the list of contributors for John Ogilby's *America* also included a Good Samaritan by the name of Gary Menges, who wrote a personal check for fifty dollars. When I asked him about this, he was dismissive. "I always contribute to the Save a Book," he simply explained. "I think it's an important program."

After the money had been raised, *America* was tenderly wrapped up and sent to the unlikely destination of Browns Summit, North Carolina, where it was unpacked with equal care. Gary Menges could not fight his war against oblivion alone; his efforts now hinged on the expertise of Don Etherington, a Book Keeper of a very high order. For the past fifty years the British-born Etherington has worked as a conservator, restoring countless old volumes with the steady hands of a surgeon and the mystic passion of a necromancer. This unlikely career began when, at age thirteen, Etherington enrolled in a special book-binding curriculum at London's Central School of Arts and Crafts. What on earth would draw someone to such an arcane endeavor at so young an age? "I have no clue," Etherington told me with a laugh, during a phone interview. Perhaps it was simply in his blood. "My uncle was the head of the reading room at the British Library for many years," he said. "And in fact, as strange as it may seem, a few years ago I found out that my great-great-grandfather was the painting restorer for Queen Victoria."

Beginning with a seven-year apprenticeship at Harrison's & Sons, the printer for the royal family, Etherington worked his way to the top of his profession in England. In 1970 he came to the United States to help set up a conservation program at the Library of Congress, then took a similar position at the University of Texas in 1980. Since 1987 he has been president of the for-profit Etherington Conservation Center, a division of Information Conservation, Inc. During the course of his career, Etherington has worked on some of the Western world's most important documents. While at the Library of Congress, for example, he helped restore the Gettysburg Address. He has also done work on the Magna Carta of 1297 (owned by Ross Perot), the Carolina Charter of 1663, the Texas Declaration of Independence of 1836, and the State of Virginia's copy of the Bill of Rights. At the time I interviewed him—September 1998—Etherington was part of a team designing a new housing and display for the Declaration of Independence.

Working on such monumental—not to mention fragile—documents requires both a lot of talent and a lot of nerve. "You cannot allow emotion to get into it," Etherington explained. "You can't be overawed by the object. You have to be concerned about what you're doing and what's wrong with the object and what's needed to restore it. If you approached it as if you were scared of it, your hands wouldn't be so sure. And that's when all the accidents happen. You've got to have a certain surety of hand to deal with this type of material. And confidence—but not overconfidence."

You've also got to have decades of training. No university of-

fers a specific major in book conservation, explained Etherington. Instead, it is largely "an artisan's job"—one of the last remaining professions in which skills are passed down from one
generation to the next. He employs a wider variety of materials
than his forebears did, but otherwise "the same procedures that
they used in the fifth or sixth or seventh century we still use
today."

With the book from Menges, Etherington's approach was
the same as it has been on thousands of other old texts. The
overall goal was to save as much of the old material as possible.
In this case, however, some elements were beyond hope. "The
binding had deteriorated and we couldn't restore it," said Etherington. "So that's the first decision: All right, we can't save this
binding, so we're going to rebind it. And then, what you do is
take off the old binding and clean the old spine. And then you
go to the text block to see if it is sound—to see whether it needs
restoring or whatever—because that's a major decision. If you
get into taking the text block apart, it's what we call major intervention, because you are getting into the original structure
and maybe changing it. Minimal intervention is when you don't
do too much to change the original configuration of the book."

Only minimal intervention was needed on the Ogilby. Other
than some torn maps, which had to be carefully patched with
material known as Japanese paper, the text block did not need
much work. Etherington and his staff could focus on the binding. "A new binding generally has new boards and new endpapers," he said. "Endpapers are the papers you put on either side
of the text block. They're the connectors between the cover

and the book. And then you decide what kind of cover you're going to put on the text block."

It was decided to use a leather binding that—while not as ornately decorated as the one that had preceded it—would be both faithful to the era in which the book was published and durable enough to be used widely by library patrons. But even such a no-frills cover required meticulous craftsmanship. "You have to pare the leather, which means you thin it down on the edges," he explained. "Then you paste the leather out, and then you form the leather around the boards of the book. You turn the leather in and then you let it dry, and once you've done that you put the endpapers down."

Once the binding was completed, the book was ready for shipment back to Seattle. Even though it had not been Etherington's most high-profile assignment, there was, as always, "that satisfaction of saving something from deterioration or the trash," he said. "And that's the difference between book conservators and other kinds of conservators. We're actually restoring objects to be reused, not just to be put in a museum . . . which I think is great. I don't think we should be restoring things to put in a glass case, never to be handled. Books were never designed to not be read."

And this one *would* be read, at long last. The cataloging process having been completed, *America* was made available to patrons of the Allen Library in June 1995. Gary Menges's effort to get it on the shelves had taken five years.

Perhaps you've already guessed what happened next. If so, it won't surprise you that on October 4, 1995, a stranger walked into the Special Collections and Preservation Division and approached the reference desk, where Menges was working. Nor will you be shocked that the man was wearing a blue blazer and, according to Menges, "looked very normal. He could have been a faculty member or a book collector. He wasn't anyone that you would be suspicious of from his appearance or the way he acted." Perhaps you've figured out by now that the man filled out a call slip for John Ogilby's *America* and that, as Menges recalled, "I told him that it was a wonderful book and went and got it for him." And if you've deduced all of this, then you need not be told that the stranger had identified himself as James Perry. There may be only one dark little detail you don't already know: he was the very first patron to handle the book.

On the day I interviewed Gary Menges at his office, he calmly recounted the story of the intruder's visit to the library, then stood and led me over to the book. Its thick new leather binding was magnificent, but beneath the cover much was amiss. One of the four maps the thief had sliced from the book was still missing, apparently gone forever. The three others had, in the intervening months, been returned by law enforcement officials and now lay next to the volume. They could be reattached—perhaps even well enough so that a casual observer might not notice any damage—but the book's integrity and monetary value had been irreparably harmed. That was especially true because a number of additional pages had been left half-mangled by deep slashes along the fold. "See how he cut the title page here?" Menges said, gesturing to one razor-scarred page. "He would cut very fast when somebody wasn't looking. And in doing so, he would cut through other pages like this one."

I had seen similarly mutilated books at other libraries but still found the sight chilling. And I wondered how it made someone like Menges feel. The conservator Don Etherington once confessed that it pains him to see even a cheap paperback ill-treated. "When you spend a whole life doing the work we do, you don't even like to see people turning the corners of books," he said. "You know how people turn in the corners to mark something when they're reading? Well, it gives you a little gut check. You feel like saying to the person, 'I wish you wouldn't do that.' That sounds a bit stupid, really—but that's the way you are. If you're in this business, you have that innate sense that you don't want these things to be mishandled."

And as for someone who would destroy a rare and beautiful book like the Ogilby? "You'd like to cut his balls off, basically," said Etherington.

As I watched Menges, however, I could see none of Etherington's ire. He looked concerned, to be sure, but calm. Nor had I perceived any great wrath during a phone interview a few days before my visit to Seattle, when I asked the librarian to describe his feelings about the theft. "Of course, one is angry," he said, choosing each word carefully. "You put personal effort into obtaining a book for the collection and getting it conserved and making it available to people. And then the first person who uses it comes in and cuts out maps. Obviously, one is not very happy about that."

By coming here, I had thought that I might get a more impassioned reaction. In the smarmy theatrics of my imagination, I had conjured up scenes in which Menges prostrated himself over the book, cursing the Library Gods and raining tears down upon the ruins of the text. Now, however, I realized just how silly such expectations had been. Gary Menges was not mourning the past. He was already looking to the future, telling me about his plans for repairing the book. He had saved it from oblivion once, and would do so again. The Keeper of the Books must never rest.

OW BIG IS THE WORLD? FOR ANYONE INTERESTED IN mapmaking, that question is about as fundamental as they come. The first person to make a scientific measurement of the Earth's circumference, however, was not a cartographer but a librarian.

His name was Eratosthenes, and, from about 235 B.C. until about 195 B.C., he was head of the library at Alexandria, one of the most legendary sites in all of antiquity. Alexandria was then the largest city in the world, a hub of Hellenistic trade and learning. The city had been founded around 332 B.C. by Alexander the Great; the library was established a few decades later—and quickly set about accumulating texts the way Alexander had acquired territory. Agents fanned out across the known world with orders to bring back the works of what one contemporary observer described as "poets and prose-writers, rhetoricians and sophists, doctors and soothsayers, historians and all the others too." Scholars were brought in to translate the texts of cultures from Europe to North Africa and India. Ships arriving in Alexandria's harbor were forced to lend the library any books they might have on board. Copies of the books—made on cheap papyrus—were eventually returned to the owners, but the originals remained in Alexandria. The library's goal was audacious: to gather, under one roof, every book ever written.

It never happened. Although, at its height, the great library built a collection of as many as seven hundred thousand rolls of papyrus, it was doomed to an inauspicious fate. No one can be precisely sure what became of it. Leading theories have the library being (a) burned by Julius Caesar in the first century B.C., (b) gutted by early Christian fanatics, (c) destroyed by Arab invaders in the seventh century A.D., (d) wasted slowly by the forces of time, or (e) all of the above. At any rate, the library vanished—but not before it helped the rest of the world to become visible. Claudius Ptolemy apparently studied there in preparation for his landmark *Geographia*. So did another prominent Greek geographer, Strabo, whose insistence that "it is possible to sail round the inhabited world on both sides, from the east as well as the west," inspired Columbus and other explorers to seek out new routes to the Indies. And then there was Eratosthenes. Thanks to a combination of brilliant abstract thinking and practical geometry (not to mention a bit of luck), he was able to deduce the Earth's circumference by measuring the shadow of a single obelisk. His estimate is hard to pin down because of our imprecise understanding of ancient Greek units of measurement. But modern scholars agree that his guess was astonishingly close to the mark. In fact, "Eratosthenes' measurement may have been within two hundred miles of the correct figure of the circumference of the earth," wrote the map historian Norman J. W. Thrower.

In those days, size mattered. Or, as the philosopher Hannah Arendt wrote in her 1958 work, *The Human Condition,* "distance ruled." It was perhaps the single most important issue for every cartographer and explorer until as late as the nineteenth century. Columbus, for example, believed that the distance between the Canary Islands, off the northwest coast of Africa, and Japan was only 2,760 miles. He might never have set sail—or been allowed to—had he known that his estimate was short by nearly 9,500 miles. And Magellan? He "had not the remotest idea of the width of the Pacific Ocean, uncrossed as yet by any European," wrote the historian Samuel Eliot Morison, who concluded that all authoritative estimates Magellan could have known about were at least 80 percent short of the actual distance. The captain and his crew were so ill-prepared for the vastness of the Pacific that during their one-hundred-day crossing they were forced to eat rats, leather, and sawdust.

In overcoming such hardships, those two journeys accomplished the miraculous: they made the world bigger. And for hundreds of years it continued to grow. But then—once the map was finally filled in—a curious thing happened: the world started getting smaller again. In *The Human Condition,* Arendt described "the shrinkage of space and the abolition of distance through railroads, steamships, and airplanes." And since her time the withering of the world has become even more profound. On the desk in front of me now is a book called *The Death of Distance*. Its bold thesis is summarized on the dust jacket: "Geography, borders, time zones—all are rapidly becoming irrelevant . . . courtesy of the communications revolution."

Geography *irrelevant*? Columbus and Magellan would have found such a concept both absurd and depressing. Yet I have to concede that Frances Cairncross, the author of *The Death of Distance*, makes a fairly convincing argument about how new telephone, television, and computer technologies "will help to shrink the world and to make people realize the extent to which, in John Donne's words, 'No man is an island, entire of itself; every man is a piece of the continent, a part of the main.' "

As I was reading *The Death of Distance,* it occurred to me that there are now two types of adventurers. The old breed—the kind who once slogged forth into the great unknown, back when there was enough unknown left to accurately describe it as "great"—is an endangered species, and not a happy one, either. The old breed does not want to be part of the main. It yearns for islands. It feels grounded in the global village. As the mountain climber Gaston Rébuffat once put it, "In this modern age, very little remains that is real: night has been banished, so have the cold, the wind, and the stars." And so the old breed finds itself jammed into the last fragments of true man versus nature wilderness, adventure ghettos like Mount Everest.

The new breed talks much the same game as the old, appropriating Age of Discovery language to describe Age of Information concepts (Netscape Navigator, Microsoft Explorer). But the similarities end there. The new breed has no need for physical wilderness. It celebrates the fact that, as Rébuffat put it, very little remains that is real. The new breed dances on the grave of poor old Distance, believing that in cyberspace all vistas are endless. At the beginning of the twentieth century, Theodore Roosevelt, one of the last great icons of the old breed, argued that the adventurer's heart "must thrill for the saddle and not the hearthstone." At the beginning of the twenty-first century, the new breed of adventurer must thrill only for the placid glow of a computer screen.

In his own curious quest, Gilbert Bland seems to have had in common with the old breed a certain compulsion toward risk. But for the most part he was solidly of the new breed. In the early 1990s, before turning his interest to maps, Bland ran a computer consulting firm, and he apparently put his technical knowledge to good use during his crime spree. He did not make his actual conquests on the Web, of course. But, according to law enforcement officials, he did do his exploration and discovery there—using the Internet, for example, to track down Ogilby's *America* at the University of Washington. Gary Menges found some bitter irony in that. "You tell the world you have something in order that people are aware of its existence and can come and use it," he said. "And then you have people who are not using this information for scholarly purposes. They're using it to put together their hit lists."

But if the death of distance helped Bland to stalk libraries, it also helped the libraries hunt down Bland. In the past, word of his apprehension in Baltimore would have traveled slowly, through gossip and other largely informal channels. But thanks to Cynthia Requardt's posting on ExLibris—a news group established in 1990 for discussion of rare books and manuscripts librarianship—institutions all over the country received notice of the crime only hours after it had taken place. "That made it a whole lot easier," Requardt later remembered. "If I didn't have ExLibris, I wouldn't have known how to contact people. I would have just stuck with the phone calls to people at the libraries I identified from the notebook. . . . The speed with which Bland's trail unfolded was amazing."

And as the librarians began to tally their losses, a collective rage spread through their ranks. "I felt like a real victim, like it was a personal assault," said Evelyn Walker of the University of Rochester. "I've been here about fifteen years, and quite a few of my colleagues have also been here for quite a while. We feel very responsible for our collections, and very connected to them. We work hard to develop them, and feel really happy when we find just the right item to fit into a little slot. It's more than a job. It's a calling, a passion, whatever you want to call it. The librarians here really are curators who treasure their materials and want to make them available. Anybody in the world would have been welcome to come here and look at those maps and use them. And now they're gone."

Added John E. Ingram of the University of Florida, "What I find to be the most difficult part is to realize that someone was coming in and destroying part of our heritage. We are a state institution, and the person who took the maps was robbing the entire state and the country, not just the library. For instance, for one of the titles—*Modern History, or, the Present State of All Nations* by Thomas Salmon—we have the only complete copy in the state of Florida. Well, we formerly had the only complete copy."

The librarians, a legendarily docile people, now wanted blood. "If Bland gets in front of my car," said Northwestern University's Russell Maylone, "I'll run over him—but in a nice way. . . . Oh, and then I'll back over him again."

But capturing Bland again did not prove to be a simple matter. After being detained in Baltimore, he apparently stayed on for a few days in his old hometown of Columbia, Maryland. While there, he called the Peabody Institute's security chief, Donald Pfouts, to request the return of his notebook. "He said he forgot his book and he really needed to get it," explained Pfouts. "Once he found out he wasn't getting his book back, I think that's when he really realized what the possibilities were. And that's when he fled."

It would take law enforcement officials nearly a month to capture him—a costly delay, as things turned out. The problem was not that the invisible man had disappeared once more. Police knew just where to find Gilbert Bland: at his home in Coral Springs, Florida. But getting there to arrest him proved to be a more arduous journey than anyone could have imagined. Bureaucracy, indifference, and disorganization can be more difficult to navigate than the widest ocean. And sometimes the world can still seem like a very big place indeed.

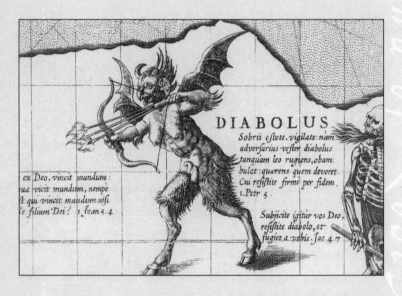

DIABOLUS

Sobrii estote, vigilate: nam adversarius vester diabolus tanquam leo rugiens, obambulat: quærens quem devoret. Cui resistite firmi per fidem. 1. Petr 5.

ex Deo, vincit mundum: ...ue vicit mundum, nempe ...t qui vincit mundum nisi ...e filium Dei? 1. Ioan 5. 4

Subijcite igitur vos Deo, resistite diabolo, et fugiet a vobis. Iac. 4. 7

DETAIL FROM A WORLD MAP BY
JODOCUS HONDIUS, CIRCA 1597,
WARNING AGAINST THE TEMPTATIONS
OF SATAN. "BE SOBER, BE VIGI-
LANT," READS THE ACCOMPANYING
BIBLICAL TEXT, "BECAUSE YOUR
ADVERSARY THE DEVIL, AS A ROARING
LION, WALKETH ABOUT, SEEKING
WHOM HE MAY DEVOUR."

CHAPTER 7

A Brief History of
Cartographic Crime

So THERE I WAS, STANDING IN AWE BEFORE the "monster camera," as one of its operators called it, a contraption of such massive proportions that it stretched across two large rooms, was able to snap a photo negative four feet wide and seven and a half feet long, could blow up an image to almost five hundred times its original size, and could reduce it by the same proportion. With one blink of its big reptilian eye, it could have transformed me from a six-foot-two-inch man into a half-mile-high colossus or a microorganism of about one eighth of an inch, just as it had once reduced all of Europe to the size of a postage stamp. It was a technological wonder, that camera. It was also, according to Vera Benson, a "white elephant," destined for the trash heap.

Benson is director of cartography for the American Map Corporation, part of the Langenscheidt Publishing Group, which competes with Rand McNally for retail primacy in the U.S. map market. A self-possessed woman with a wry sense of humor and a faint German accent, Benson was giving me a tour of the company's headquarters, a no-frills brick building in an industrial area of Queens, New York. If I had been following directly in the path of Gilbert Bland, this stop would not have been on my itinerary. To the best of my knowledge Bland neither came here nor had anything to do with American Map—unless, of course, he used one of the firm's popular road atlases to find his way from one library to the next. But in my journalistic travels, as in my personal wanderings, I'm a sucker for detours, back roads, tourist traps, scenic views, and historic landmarks. Invariably, these side trips enrich the journey, and every now and then they lead to an important discovery. I had come to Queens out of simple curiosity. Having spent so much time studying the profession's previous thousand years, I wanted to find out about the state of commercial mapmaking at the edge of the next millennium.

"We have traditional cartography and digital cartography going on down here," Benson explained, showing me into the mapmaking department, a long, open basement room with low ceilings. But even if she hadn't mentioned it, I could have seen that American Map was in a state of technological transition. One part of the room was filled with drafting tables, strewn with art supplies and staffed by slightly harried-looking designers. They were using a mapmaking process that was cutting edge not too long ago but that now struck even a casual observer like me as slow and inefficient—involving, among many other maddening details, the manual application of literally tens of thousands of street names to a page. American Map's colossal camera, that fading marvel of the mechanical age, was the central component in this kind of cartography, producing the negatives used to print everything from giant wall maps of the world to wallet-sized street maps of midtown Manhattan.

Another part of the room had a much more orderly appearance. Here, the tabletops were tidy and the faces serene. Instead of grappling with X-Acto knives, scribing tools, pens, paper, and unwieldy sheets of Mylar, the cartographers were drafting maps within the clean confines of their computer screens. "The advantage obviously is that digital maps are much easier to revise," Benson told me. "They're very time-consuming to create, unlike what people might expect. But then the benefits come in when you want to change the look of a map, eliminate features, alter it. Each feature is coded, so that, at will, you can just remove layers of information. You can take out a certain size of type, for instance, or you can change all the yellow roads to green. Previously, if you made a whole map and then decided the type you used was much too bold, there was nothing else to do but to scrap the whole thing and add on another fifteen thousand names by hand. Now, all you do is push a button."

It's not always quite that easy, of course. In another part of the room, Benson introduced me to a frustrated cartographer who was "looking at the fallout" from a computer-generated atlas that encompassed Bergen and Passaic Counties in New Jersey and Rockland County in New York. "We just got the index, and it has a lot of mistakes," Benson said. "There are twelve thousand street entries just for Bergen—and twenty-three hundred discrepancies. . . . Somehow the computer manages to create more problems on top of the ones we already have. We can't figure out why."

Because of such technical problems and other practical concerns, Benson predicted it would be years before the giant camera would snap its last shot. In the meantime, even workers who make maps the old-fashioned way were undergoing extensive computer training as the firm rushed to embrace the digital age.

Yet despite all the changes at American Map, what struck me most about the place was a sense not of disruption but of continuity. As Benson led me around the facility, I often felt like I was looking into the past, present, and future of mapmaking all at the same time. The decor, for example, mixed satellite photos of the Earth with reproductions of maps from the seventeenth and eighteenth centuries. And, glancing around the cartography department, I was amused to observe a state-of-the-art computer on one desk and a stack of books about the history of cartography on another. The texts were there, it turned out, because the firm was trying to cash in on the old maps craze. "We are turning one of our world maps into an antique-look map," Benson explained. "It's going to be an up-to-date world map, but it's going to look like an antique map, designed for those people who might not want some stark modern map in their office but may want something more traditional."

I wondered whether this computer-designed artifact would include sea monsters.

"Absolutely!"

Though she professed no expertise in the subject, Benson had an obvious appreciation for old maps. On the wall of her office hung a huge reproduction showing a bird's-eye view of Manhattan from the nineteenth century, and behind her desk stood a framed engraving of Johann Baptist Homann, a famous German mapmaker of the 1700s. "A friend of mine who is an antiques dealer thought I might like to have [Homann], because he was a German cartographer and I am a German cartographer," she explained with a shrug. "I had never heard of him."

The two mapmakers certainly *looked* nothing alike. The man on the engraving had a tight-lipped expression every bit as pompous as his lace cravat and cuffs, and wore a full-bottom wig so absurdly overblown that it looked like a sheep had crash-landed on his skull. Benson, by contrast, had short brown hair, sensible glasses, and simple elegance to go with her direct, low-key manner. Yet the more I listened to her talk, the more she seemed to have in common with Homann and other mapmakers from the past. Benson began her work life in Germany as an architectural draftsman and later studied to become a civil engineer. Those plans were cut short, however, when she married an American and moved to the United States. She then worked in the fine arts—"graphics, painting, lithography, photography, and the like"—before coming to American Map two decades ago. She freely conceded she was not a "classically trained cartographer" but made no apologies for it: "Map publishing, which is what we do here, is really an interdisciplinary field that combines cartography, printing, art, photography."

The man in the picture, it turned out, had not been classically trained, either. Although late in life Homann was given the title Geographer to the Holy Roman Emperor, he began his career as an ordinary engraver. Like Benson, he succeeded because he knew how to merge science and commerce and art, how to bring the physical world to the printed page in a way that was both useful and beautiful. The technology of publishing is now vastly different than it was in Homann's time, but the instincts and imagination required to make maps have not changed—nor, as Benson pointed out, has the fundamental objective: "information, how to get it right and get it in a timely way. . . . In cartography, that's the big word: new data."

New data was what had driven Ptolemy and Ortelius and Homann, just as it was what now drove Benson and her contemporaries. But if the quest to procure new data had remained constant, so had the need to protect it. Before I contacted American Map, Rand McNally had refused my request for a tour of its plant; a publicist hinted that the firm was worried about the loss of trade secrets. Benson was obviously less guarded, but she made it clear that security was also an issue for her company. American Map, she explained, uses a trick called trap streets to discourage competitors from infringing on its copyright. This practice, common to the industry, involves hiding a fictional roadway somewhere on each map. "We place the trap streets in areas that would be relatively harmless and would not mislead someone using the map—just a cul-de-sac at the end of some development," Benson said. "I let my researchers be creative in deciding on the names: they might name the street after their wife or dog or whatever. This allows us to do a quick spot check of our competitors' maps to see if they have stepped on our toes. It happens all the time.... Sometimes it's an innocent blunder, but more often, it's not so innocent."

And so, sitting in Benson's office and imagining maps with out-of-the-way streets such as Fido Lane or Fritzie the Good Puppy Boulevard, I realized that yet another big factor had remained constant throughout the history of cartography: larceny.

CALL ME OBSESSED, BUT THE MORE I LOOKED INTO MAP theft, the more I found it lurking behind the scenes during key moments of our collective past. And I'm not talking about *small* moments, either—unless you consider Columbus's voyage to America small or Magellan's circumnavigation of the globe small or the invasion of Normandy small. I started to think the historians had overlooked something important in their eternal search for those "hidden hands" that guide the course of human events. I kept coming upon hidden hands myself, all of them holding hot maps. It was not that I presumed to have hit on the answer to all life's mysteries, a Unified Theory of the Missing Map. It was just that I began to see Gilbert Bland as heir to an ancient though hardly noble tradition, one that has shaped our world more than we know.

Maps are big-time booty. Mapmaking has been going on since at least the Stone Age: a picture map painted on a wall of the famous Catal Huyuk settlement in south-central Turkey dates back to about 6200 B.C., nearly three thousand years before the first system of written language appeared. And recent studies have shown that even very young children have an uncanny ability to understand maps, fueling speculation that we are born with the skill. All cultures are thought to make maps in one form or another, and for good reason. Throughout history they have enabled their possessors to win military battles, gain access to precious economic resources, and claim new territories as their own.

As manna, maps predate even money. Cartographic crime also goes back thousands of years. "In the Roman Empire . . . maps of the world were exclusively for government use, and it was a crime for a private person to possess one," wrote the historian Daniel J. Boorstin. Augustus, emperor at the time of Jesus' birth, was so worried his maps of his empire might fall into the wrong hands that he had them locked in the innermost vaults of the palace. Rome's traditional rivals from Carthage were no less concerned about theft. According to legend, one Carthaginian sea captain sank his ship rather than let his sea charts fall into Roman hands. His crew drowned, but he was given a hero's welcome upon his return home.

In the Age of Discovery map theft literally changed the world. Sooner or later some European adventurer would have stumbled upon America, but it might not have been Christopher Columbus and it might not have happened in 1492 if not for some sticky fingers. Before beginning work for the Spanish crown, Columbus spent the better part of nine years in Portugal, where his brother Bartholomeo had established himself as a noted cartographer. In 1485, having failed to sell Portugal's King John II on his proposed Enterprise of the Indies, Columbus decided to solicit help for the plan in Spain. He left in secret, "fearing the king would send after him and hold him," according to the sixteenth-century chronicler Bartolomé de Las Casas. The reason for his furtive exit is unclear—but some historians have wondered whether Columbus absconded with a copy of the Florentine cosmographer Paolo del Pozzo Toscanelli's world map (no longer extant), which showed a possible westerly sea route to the Indies. If so, the mariner would have had good reason to worry about being stopped. Portugal was then the world's dominant seafaring nation, a status it guarded with an iron fist. Nautical writings and charts were considered state secrets, and to copy or divulge their contents was a capital crime. Yet despite these restrictions, an even bigger cartographic caper would soon follow. This time the perpetrator was Bartholomeo Columbus, who had remained behind in Lisbon to help compile a large world map that incorporated the latest top-secret information gleaned by Portuguese explorers. As Lisa Jardine explained in *Worldly Goods: A New History of the Renaissance:*

> Bartholomeo prepared to join [a cash-strapped Christopher Columbus] in Spain to help his project. Before he left Lisbon, however, he copied a number of Portuguese

maps from the secret archive, including the large world map (which he transferred on to eleven sheets of paper, because paper was lighter, and easier to conceal, than parchment). The Columbus brothers sold Bartholomeo's valuable stolen maps in Italy for substantial sums, thereby dispersing vital information hitherto held only by the Portuguese.

In Seville, their financial circumstances alleviated by their map sales, the two brothers reassembled the world map, and modified it to incorporate material not yet entered at the time of Bartholomeo's departure. Crucially, this additional information suggested that, once the Cape of Good Hope had been rounded, a significant mass of land remained still to be navigated around before ships could gain access to the Indian Ocean. Although factually incorrect, the Columbus brothers' map, thus modified, apparently offered strong concrete support for the argument that a westwards route to the Indies was a viable alternative option. On the strength of comparison of the Columbuses' map (with its exaggerated coastline of Africa) with their own Ptolemaic maps (on which calculation errors significantly reduced the Atlantic Ocean) the Spanish monarchs Ferdinand and Isabella finally agreed to finance the venture in 1492.

All this from a purloined map. No wonder Lisbon kept a tight lid on its secrets—so tight, in fact, that almost none of the empire's charts made before the sixteenth century have survived, meaning that much of what we know about Portuguese cartography comes from the relatively few maps that left the country illegally. The most famous of those is the so-called Cantino planisphere, a one-of-a-kind world map with a colorful and iniquitous history. Considered the earliest surviving map of Lisbon's discoveries east and west (and one of the first maps ever to show the Americas), the Cantino planisphere was named not after the unknown Portuguese cartographer who drafted it but after the Italian secret agent who smuggled it out in 1502. Alberto Cantino had gone to Lisbon masquerading as a dealer in purebred horses. His real mission, however, was to gather intelligence for the duke of Ferrara. Gather it he did, bribing a mapmaker for an up-to-the-moment copy of the *padrão real,* the standard cartographic prototype unto which new discoveries were constantly being added.

Prominent on this lavishly illustrated chart is a fragmentary outline of the Brazilian coast, marked by Portuguese flags and decorated with resplendent red macaws. Just two years earlier the adventurer Pedro Álvares Cabral had sighted this new land, which he "believed to be a continent," according to an inscription on the map. That was big news. Christopher Columbus went to his death in 1506 insisting that the land he had reached was Asia, but other European navigators were beginning to realize that the face of the Earth was drastically different from the three-continent version pictured on their Ptolemaic maps. This was one of the most sudden and dramatic jolts in the history of human thought, and it triggered a fierce competition for cutting-edge cartographic data, which, given the realities of the time, often meant contraband cartographic data.

Among those exploiting this emerging black market was Ferdinand Magellan, who departed his native Portugal for Spain in 1517, taking with him classified information that a navigable strait might exist at the extreme southern end of South America. At a meeting in Spain with Holy Roman Emperor Charles V, Magellan "brought with him a well-painted globe showing the entire world," wrote Las Casas, "and thereon traced the course he proposed to take, save that the Strait was purposely left blank so that nobody could anticipate him."

Scholars disagree about the origin of this globe. Some have argued that it was either a copy or an updated version of a famous 1492 work by Martin Behaim, today considered to be the oldest surviving European terrestrial sphere. Behaim was an enigmatic mathematician, merchant, and adventurer who had once been a government mapmaker in Portugal but had since sold his services to the city fathers of Nuremberg. "In making the globe for the Germans," explained Lisa Jardine, "Behaim had been involved in a substantial piece of commercial and industrial espionage. . . . Behaim incorporated the cartographical information he had access to in Lisbon—the most highly classified and inaccessible cartographical information currently available."

Other historians, however, have conjectured that what Las Casas loosely described as a "globe" was, in fact, a world map— most likely one made by the Portuguese cartographer Jorge Reinel, who had gotten into trouble in Lisbon and had since fled to Spain. This argument is bolstered by a report, made in 1519 by Lisbon's ambassador to Seville, which informed the Portuguese king that Reinel had made a map specifically to help Magellan prepare for his voyage. Reinel's father, the prominent mapmaker Pedro Reinel, had gone to Seville to bring his wayward son home. But according to the ambassador, the elder Reinel was also seen adding details to the Spanish map.

In any case, it is clear that Magellan benefited from cartographic secrets stolen from the Portuguese. A firsthand account of his journey suggests that without access to such intelligence, he might never have located the famous strait that now bears his name. On October 21, 1520—more than a year after setting sail—he reached a cape where the South American coastline appeared to turn west. Was this, at last, the long-sought opening to the Pacific? Antonio Pigafetta, an Italian who sailed with the expedition and later wrote about it, was not convinced. Like others on the journey, he had expected the strait to be an open passageway that you could see through, like the Strait of Gibraltar. This cape, "closed on all sides," did not inspire much hope. But Magellan was certain they were in the right place—apparently for good reason. As Pigafetta explained, "He knew where to sail to find a well-hidden strait, which he saw on a map in the treasury of the king of Portugal."

Historians are still unsure which map Magellan had seen, or whether it actually could have provided him much useful information. And no map could have guided the ships through the unexplored strait that lay ahead, a maze of narrow passages and small islands, blasted by strange winds. Magellan's skills as a navigator should not be underestimated. Still, without the illicit information he procured from Lisbon, the first circumnavigation of the Earth might have been neither attempted nor completed.

But the Portuguese were not alone in their worries about map theft. Spain was quickly establishing a trade monopoly in the New World that rivaled the Portuguese hold on Africa and Asia. And as Daniel Boorstin noted in *The Discoverers,* the new empire had security problems of its own:

> The Spanish . . . kept their official charts in a lockbox secured with two locks and two keys, one held by the pilot-major (Amerigo Vespucci was the first), the other by the cosmographer-major. Fearing that their official maps would be deliberately corrupted or would not include the latest authentic information, in 1508 the government created a master chart, the *Padrón Real,* to be supervised by a commission of the ablest pilots. But all these precautions were not enough. The Venetian-born Sebastian Cabot . . . while serving as pilot-major to Emperor Charles V, tried to sell "The Secret of the Strait" both to Venice and to England.

Perhaps no one person was more hated by the Spanish than the man they called "master thief of the unknown world." In his fabled round-the-world voyage of 1577–1580, Francis Drake not only helped shatter Spain's hold on the Americas but came back to England with some forty tons of gold and silver bullion from Spanish ships and outposts. And yet those were not his only spoils. On March 20, 1579, while storming a small enemy vessel off Costa Rica, his men captured two large navigator's maps of the Pacific and a collection of charts detailing Spain's China route. Nuño da Silva, a Portuguese pilot Drake was then holding captive, later testified that the English buccaneer "prized these greatly and rejoiced over them." Drake had good reason to be pleased. Many historians believe that he had not left England with firm plans to circle the world. But now, with Spanish ships searching for him up and down the Pacific coast of South America, he was in no position to head back through the Strait of Magellan. The charts of the Pacific were "exactly what he required," wrote Alexander McKee in *The Queen's Corsair: Drake's Journey of Circumnavigation, 1577–1580*. Added the historian John Hampden, "Drake's smooth passage from the American coast to the East Indies was not due solely to his genius for navigation; he had [the] captured Spanish charts."

Other swashbucklers quested after similar booty, as Lloyd A. Brown observed in *The Story of Maps*:

In the sixteenth century genuine Spanish charts of any part of the Americas were real maritime prizes, rated as highly by the French and English as the gold bullion which might be in the ships' strong rooms. One such priceless haul was made by the English adventurer and freebooter Woodes Rogers. While cruising on behalf of some merchants of Bristol along the coast of Peru and

Chile he captured some charts which were so "hot" that they were immediately engraved in London and published by John Senex.

In another celebrated case, a group of English pirates led by Bartholomew Sharp captured the Spanish ship *Rosario* off the Ecuadoran coast in 1681. On board they found what one buccaneer described as "a great Book full of Sea-Charts and Maps, containing a very accurate and exact description of all the Ports, Soundings, Creeks, Rivers, Capes, and Coasts belonging to the South Sea, and all the Navigations usually performed by the *Spaniards* in that Ocean." A Spanish mariner tried to cast this chart book—or *derrotero*—into the sea, but one of the pirates stopped him. "The Spaniards cried when I gott [*sic*] the book," Sharp later wrote.

Sharp did not know it at the time, but that *derrotero* may have saved him from the gallows upon his return home in 1682. Because England and Spain were then at peace, the buccaneer and two members of his crew were charged, at the behest of the Spanish ambassador, with piracy and murder. Before the trial, however, a copy of the stolen chart book was given to England's King Charles II, who took a personal interest in the case. Sharp and the others were soon acquitted—a verdict that may well have been due to "royal influence," according to Derek Howse and Norman J. W. Thrower, authors of *A Buccaneer's Atlas,* a book about the incident.

Charles II, by the way, seems to have had a real passion for the cloak-and-dagger side of cartography. In 1683 England's ambassador to France wrote the king to offer "the plans of some fortified places which I believe are very exact. If Your Majesty . . . approve of them, I do not question but to have draughts of all the other fortification of France in a little time." England and France were then allies. No matter: a reply soon came back informing the ambassador that Charles had received the plans and "carried them to his closet: where . . . he perused them . . . appeared extremely well pleased and . . . desire[d] you would goe on & procure as many as you could."

Another leader who appreciated a good secondhand map was George Washington. A former surveyor, the general was well aware of the military necessity of knowing one's terrain. In 1777 he lost more than twelve hundred men at the Battle of the Brandywine, partly because he had no accurate map of the area. "Spies directed by Washington were explicitly alerted to acquire maps whenever possible," wrote the map historian J. B. Harley, who added that "men were . . . prepared to risk their lives (and lost them) in acquiring suitable maps."

The crown had its own spies. Benedict Arnold's scheme to join the British was only discovered when his illicit maps—the plans for West Point—were found hidden in the boot of an accomplice. Other map-toting turncoats had better luck. In 1776 the British captured Fort Washington in New York City, thanks at least in part to the treachery of an American officer named William Demont, who later confessed that he had "brought in with me [to the British] the plans of Fort Washington, by which plans that fortress was taken by His Majesty's troops."

Stolen maps were again at a premium in the Civil War—

especially for the Confederacy, which generally lacked not only trained cartographers but mapmaking basics, such as surveying equipment, printing presses, paper, and ink. At times the situation was so bad that one Southern general declared his field officers "knew no more about the [local] topography of the country than they did about Central Africa." Southern spies coveted cartographic intelligence—and often had to look no further than Northern newspapers, whose accounts were regularly accompanied by battlefield maps. "I *know* that the principal northern papers reach the enemy regularly & promptly," fumed General William Tecumseh Sherman, who added that the problem had "brought our country to the brink of ruin."

Some Confederate operatives, however, had more prurient methods of obtaining maps. Rose O'Neal Greenhow was a beautiful and well-connected widow whose admirers included a number of powerful Northern politicians and military officers. From her richly appointed house in Washington, D.C., Greenhow ran a remarkably successful espionage operation—at one point charming a U.S. senator out of key military secrets that helped the South win the Battle of Bull Run. The person who finally exposed her as a spy was Allan Pinkerton, the famous private detective. On August 21, 1861, Pinkerton—perched on the shoulder of two assistants—watched through Greenhow's parlor window as the sultry secret agent entertained a young Union officer. As Pinkerton looked on, the soldier handed her a military map and described its contents in detail. Then, according to the detective's account of the event, Greenhow led the officer into another room—from which they returned "arm and arm" an hour later. "There was no doubt, if Pinkerton's report was to be believed, that the handsome widow had performed a sexual tryst with the officer in return for his delivery of the vital map," observed the writer Jay Robert Nash.

During World War I both sides were so concerned about secrecy that their maps of the front showed only the enemy's trenches, a practice that prompted troops to capture maps from the other side in order to get around their own fortifications. World War II military planners were also obsessed with safeguarding the maps of their defenses—sometimes in vain. In 1943 Nazi officials in the French town of Caen hired a man named René Duchez to redecorate their headquarters. While there Duchez noticed that, folded on the top of a desk, was a large map labeled "Special Blueprint—Top Secret." A normal housepainter would have minded his own business, but Duchez happened to have a sideline career with the French resistance. While the German commandant was distracted by other matters, Duchez hid the map within the office, retrieving it when he returned to work a few days later. The stolen map was a remarkable find, detailing the Germans' secret "Atlantic Wall" defense system, an elaborate network of underwater obstacles, tank traps, minefields, barbed wire, and gun installations— information that proved essential to the Allied invasion of Normandy. As General Omar Bradley later explained: "Securing the blueprint of the German Atlantic Wall was an incredible feat— so valuable that the landing operation succeeded with a minimum loss of men and material."

Not every map heist of World War II worked out so well, however. In 1944 twenty-five German POWs pulled off the largest and most spectacular escape from an American compound during the war, digging a 178-foot tunnel out of the Navy's Papago Park Prisoner of War Camp in Arizona. All of the men were eventually captured, but some remained at large for more than a month. Among the last to be brought in were three German soldiers who had based their own audacious but ill-fated escape plans on a stolen highway map of Arizona, which showed the Gila River leading to the Colorado River, which in turn led to Mexico. Devising a scheme to flee by water, they constructed a collapsible kayak under the noses of their captors, tested it in a makeshift pool within the prison compound, then snuck it out through the tunnel. Their plan was perfect—except for the map. The Gila, shown as a healthy blue waterway, turned out to be little more than a dry rut.

$\mathcal{B}$ECAUSE MAPS HAVE SO OFTEN PROVED INACCUrate, out-of-date, or just plain unavailable, military leaders have long searched for ways to get the lay of the land without them. Manned observation balloons were used for intelligence gathering as early as 1794 and played a vital role in both the Civil War and the Spanish-American War. In World War I, camera-equipped airplanes were providing information not available on old-fashioned maps, such as up-to-date images of troop movements and bomb damage. Aerial espionage continued to advance and flourish through World War II and the Cold War.

But by 1960, when Soviet forces shot down the U-2, an American ultrahigh-altitude aircraft specifically designed for spying, both superpowers were already hard at work on the next frontier of overhead observation. The space race was on, and soon the sky would be full of eyes.

Far from making maps obsolete, however, new satellite technologies have led to one of the most productive periods in the history of cartography, comparable only to the golden age of mapmaking in the sixteenth and seventeenth centuries. NASA's Landsat satellites—which, rather than taking photographs of the Earth, measure wavelengths of reflected energy—have "stood traditional cartography on its ear," according to Stephen S. Hall, author of *Mapping the Next Millennium*. "By seeing in electromagnetic increments beyond the normal range of human vision, Landsat revealed whole new worlds hidden within the folds of a familiar world we thought we knew so well." Among the many benefits of Landsat imagery has been the discovery of previously unknown lakes, a new isle (named Landsat Island) off Canada's Atlantic coast, and an uncharted reef in the Indian Ocean.

While such imagery greatly enhances maps, it won't replace them anytime soon—a point I grasped one day as I toured the Santa Barbara, California, headquarters of Map Link, Inc., the largest independent wholesale distributor of maps and atlases in North America. My host that day was the firm's publishing manager, Will Tefft, a man who regularly travels to distant corners of the Earth in search of cartographic treasure. As he led me around the facility, a cavernous former lemon-packing plant, Tefft explained that he and his associates chase a lofty goal: to stock every modern-day map by every map publisher on the planet. When a desired map cannot be obtained or does not exist, Map Link will often publish one itself. Among the firm's customers have been the Kuwaiti embassy in Washington, which purchased maps of its own country during the 1990 Iraqi invasion, and the U.S. military, which also placed large orders for maps of the region during the Gulf War.

Tefft, far more than most people, understands the importance of satellite photos in modern cartography—yet he insisted that no photograph alone can take the place of a map. "Perhaps people have been overwhelmed by movies, TV news, and all the other electronic media," he said. "There's a camera everywhere; it's just incredible. But a camera can't reveal the place names—anything that humans ascribe a word to, from Mount McKinley to the Los Angeles International Airport—and a camera can't help travelers, in terms of what they are looking at now and what they are aiming for over the horizon."

I remembered his words a short while later, when, during the height of the air war over Yugoslavia in 1999, NATO forces accidentally bombed the Chinese embassy in Belgrade, killing three people and touching off a messy diplomatic crisis. In preparing to deploy "smart bombs" for the attack, NATO commanders had made use of the most advanced satellite imagery available. What they did not have was something that could show them the correct locations of both the embassy and the intended target, the Yugoslav Federal Directorate of Supply and Procurement. The resulting tragedy was graphic proof that even in an age when satellites allow us to observe each other and our world in mind-boggling ways, there's sometimes still no substitute for a good map.

Not surprisingly, a black market for up-to-date maps still exists. In 1995, for example, a Michigan man named Bill Stewart was convicted of trying to sell classified U.S. government computer tapes containing three-dimensional maps of the Mideast. Nonetheless, in an era when you can switch on your laptop and find an adequate map of almost any place on Earth, and when, for a fee, you can buy satellite images of your own home or even spy from space on the backyard of a hated neighbor, contemporary maps have lost much of their contraband currency. These days a new kind of map thief has emerged—one who steals not to claim the future, like Columbus or Magellan, but to recapture a lost and longed-for past.

WHO COULD BE MORE TRUSTWORTHY AROUND OLD books than a couple of priests? When fathers Michael Huback and Stephen Chapo—members of the Eastern Rite of Greek Orthodox Catholicism, a Slavonic sect—began making frequent visits to Yale's Sterling Memorial Library in November 1970, they apparently received little scrutiny from staff members. The clergymen remained regulars in the stacks until March 1972, when a Chicago book dealer offered to sell Yale one of three extant copies of a 1670 atlas by the Dutch cartographer Hendrick Doncker. Library officials were shocked to discover that the sale item was, in fact, the copy they had thought to be sitting safely on their shelves. The stolen atlas was soon linked to Huback and Chapo, and when librarians did an inventory of their atlas collection they discovered that nearly two hundred other books were missing. But Yale, apparently, had not been the only victim. When FBI agents searched a New York City monastery where Huback and Chapo were living, they recovered books belonging to a number of other schools, including the University of Chicago, Dartmouth, Harvard, Notre Dame, Manhattan College, the University of Washington, and Indiana University. The motivation of the priestly pair, soon defrocked, seems to have been decidedly secular. As St. Jerome once quipped: "Avoid, as you would the plague, a clergyman who is also a man of business."

The recent annals of cartographic crime are full of such colorful tales. The offenders come from varied backgrounds and circumstances, but almost all of them have one thing in common. Like Gilbert Bland, they have all wrestled with bizarre alter egos, as if what those stolen maps showed them were secret routes to the furthest places in their hearts. Among the more notable figures in this unusual rogues' gallery are these:

- Andrew P. Antippas, a Tulane University English professor who "was regarded with that special undergraduate awe reserved for only a chosen few professors," according to a 1979 report in the New Orleans *States-Item*. A charismatic lecturer who sometimes fought back tears while reading the poetry of Keats aloud in class, Antippas was apparently held in such high esteem that the student board of the cam-

pus newspaper, *Hullabaloo,* voted not to report on his November 1978 guilty plea to charges resulting from the theft
of five antique maps—then valued at twenty thousand
dollars—from Yale University's Sterling Memorial Library.
He also admitted to stealing maps from the Newberry Library in Chicago. A map collector and part-time dealer, he
apparently succumbed to his demons. "I was there, I was
tempted. Who can explain something like this?" he said at
the time. In January 1979, upon being sentenced to a year in
prison, he declared that "I indeed ruined my life and career." But in retrospect he appears to have been overstating
the case. Today Antippas owns Barrister's Gallery, an up-
scale New Orleans establishment specializing in Southern
folk art.

• Charles Lynn Glaser, one of the most notorious Jekyll-and-
Hyde figures in cartographic crime. Glaser was a noted authority on antiquities whose 1970 book, *Engraved America,*
was once described by the trade publication *AB Bookman's
Weekly* as "the standard reference for early American
iconography." But he was also a compulsive map thief with
a criminal career that spanned three decades. Even before
his legal troubles began, Glaser exhibited a curious fascination with fraud. In his 1968 book, *Counterfeiting in America:
The History of an American Way to Wealth,* he took lengths to
praise the "few great counterfeiters . . . men of unusual
skill or cunning" who "ennobled the crime by demonstrating vision and industry." It's not clear why Glaser himself

turned to crime. He did sell the maps he stole—but when I tracked down his ex-wife, Nosta Boll Glaser, she said, "I don't think he did it for the money." Instead, she offered this theory: during the 1970s her former spouse, suffering from "delusions of grandeur," became deeply frustrated that his own map and print business floundered while a fellow Philadelphian, the loud and brash Graham Arader, thrived. "Arader was a younger guy who came on the scene like gangbusters. And I think my ex-husband couldn't quite deal with his success and his presence," she explained. "Lynn considered himself—and this is his expression and not mine—*morally superior* to Arader. And since he was morally superior he couldn't quite understand why Arader was having all this success. . . . I think he justified the thefts by telling himself he was morally superior but he wasn't being appreciated."

In July 1974 Glaser was arrested for stealing eight sixteenth-, seventeenth-, and eighteenth-century atlases—including Thomas Jefferys's famous *American Atlas* of 1776—from the map room at Dartmouth College. Sentenced to a three-to-seven-year prison term, he spent some seven months behind bars before being paroled. But that did not dissuade him from striking again. In 1978 Glaser stole two maps by the explorer Samuel de Champlain from the University of Minnesota's James Ford Bell Library. The maps—one published in 1613, the other in 1632—were each worth more than $10,000 at the time (and now might fetch upwards of $300,000 and $100,000, respectively).

Upon his guilty plea in 1982, Glaser, who also admitted stealing two maps from the Newberry Library in Chicago, was given six months in prison, ordered to pay $5,500 in restitution, and sentenced to four and a half years' probation. Yet he continued to pass himself off as a genteel bibliophile, publishing another book, *America on Paper,* in 1989. "For twenty-five years I have been handling browned bits of paper, usually bound in calf with ridged spines and gold stamping that somehow survived into the late 20th century," he wrote in that volume. "Sometimes, even if I could not read a work in the original, simply handling it made me imagine I was in touch with time."

Librarians must have fumed at the notion of a book destroyer lauding the survival of books, especially after he pleaded guilty in March 1992 to stealing a map from a 1628 edition of Sebastian Münster's famous *Cosmographia,* housed at the Free Library of Philadelphia. Less than one month later, while on probation, he was discovered again— reportedly wearing surgical gloves and carrying a hammer—in the stacks of Lehigh University. He might be out there in some rare books room today, still handling those beloved "browned bits of paper."

• Robert M. "Skeet" Willingham, Jr., librarian, author, Sunday school teacher, and city council member from Washington, Georgia, who was described by his friends and neighbors as "Mr. Straight," "a perfect gentleman," and "a very gentle fellow." Somewhere along the line, however, this exemplary citizen, the rare books curator for the University of Georgia, developed an alter ego. In the words of Katharine Leab, the editor of *American Book Prices Current* and an expert on library security: "He became his doppelgänger: the *bad* Willingham." In June 1986 a staffer at the university's Hargrett Rare Book and Manuscript Library noticed that a rare map of South Carolina was missing. Then a cursory inventory of the map collection indicated that others had been removed as well. Police—led to Willingham on a tip provided by the map mogul Graham Arader—soon searched the librarian's historic columned home and turned up seventeen framed maps, several of which were known to have been missing from the university. But that was just the tip of the iceberg. In September 1988, a jury convicted Willingham on thirteen of fourteen felony counts for the thefts of a huge cache of library property, including an eight-volume set of floral prints by Pierre-Joseph Redouté, valued at $500,000 (and now worth as much as $600,000).

• Fitzhugh Lee Opie, a descendant of the Confederate general Robert E. Lee and the legendary Virginia governor Henry "Light-Horse Harry" Lee, who liked to present himself as a guardian of history. When Virginia was considering adoption of the Martin Luther King holiday, Opie, an Alexandria bookstore owner, labeled the idea "a disgrace," arguing that the slain civil rights leader was "neither a Virginian nor a patriot." Opie's professed love of country, however, did not dissuade him from gutting the Library of

Congress over a period of ten years—visiting several times a week, according to prosecutors, to steal maps, books, and prints. The scam came to an end in March 1992, when Opie was caught with two mid-nineteenth-century Pacific Railroad maps stuffed under his sweater.

• William Charles McCallum, a Yale University and Boston College Law School graduate, whom *The Boston Globe* described as "a rising star" at the New Hampshire attorney general's office. Thin, elegant, and sophisticated, with a taste for fine art and antiques, McCallum had such impressive legal credentials that the future U.S. Supreme Court justice David Souter—then on the state Supreme Court—once personally promoted him. But this outward success, his defense team would later argue, belied McCallum's bizarre private life, full of obsessive rituals such as eating his own hair, brushing his teeth up to fifty times a day . . . and stealing just about everything he could lay his hands on, including underwear and socks from laundromats. ("He only feels comfortable in stuff that's stolen," explained his attorney.) When police finally caught up with McCallum in 1996, they found his Colonial-style home in Londonderry packed with more than a hundred thousand dollars' worth of paintings, books, furniture, and prints, most of them stolen from local libraries and universities. Prominent in the ill-gotten collection were a number of cartographic curiosities, including Thomas Moule's 1845 map of Hampshire County, England, stolen from Saint Anselm College in Manchester, New

Hampshire; Jean-Baptiste Bourguignon d'Anville's 1755 map of North America, taken from the Thayer School of Engineering at Dartmouth; and two old editions of Strabo's *Geography* belonging to Dartmouth's Baker Library.

• Last but hardly least, the mysterious Daniel Spiegelman of Yonkers, New York. Among librarians Spiegelman is remembered as one of the biggest thieves in recent history, plundering the Rare Books and Manuscripts Library at New York's Columbia University for approximately $1.3 million in books, documents, and manuscripts, including more than 250 early maps. But among conspiracy theorists, especially those of the right-wing militia variety, he is even more legendary.

This much is known about the case for certain: in 1994 the staff at the rare books room at Columbia discovered that a large number of priceless documents—everything from one-of-a-kind medieval manuscripts to letters and documents signed by various American presidents—had been stolen. The perpetrator did not operate in public but broke into the secured stacks after hours, possibly via an abandoned conveyor belt, according to the library's director, Jean Ashton. When Ashton and other Columbia authorities discovered the thefts, they immediately started contacting book and print dealers worldwide—a strategy that soon paid off. In June 1995 authorities in the Netherlands arrested Spiegelman after he tried to sell a medieval manuscript to a Utrecht dealer, who recognized the item as stolen.

Then the story really got strange. Newspapers in the Netherlands, where the suspect was being held, began reporting that Spiegelman had possible links to Timothy McVeigh and Terry Nichols, suspects in the deadly 1995 bombing of the Alfred P. Murrah Federal Building in Oklahoma City. The reports quoted a Dutch Justice Ministry official as saying that Spiegelman was "suspected of delivering weaponry to the suspects of the Oklahoma bombings." The allegation—later retracted by the Dutch government—apparently stemmed from the fact that Spiegelman was wanted in the United States on charges of illegally purchasing several pistols from gun shops in Phoenix and Scottsdale, Arizona. Dutch newspapers also reported that a safe belonging to Spiegelman in New York contained a list of up to forty names of people allegedly linked to the blast.

The U.S. Justice Department, however, repeatedly denied a connection between the two cases. And Spiegelman himself disavowed any part in the Oklahoma City bombing, telling the Dutch media that "I was involved in it as much as I was involved in the assassination of John F. Kennedy." Nonetheless, the story soon took on a life of its own, encouraged by a member of McVeigh's defense team, who declared that Spiegelman bore a noteworthy resemblance to John Doe No. 2, the mystery man long sought in connection with the bombing. The Internet was soon abuzz with innuendo, as exemplified by one widely distributed posting, titled "Thirty Questions About Oklahoma City," which appeared on websites such as the Big Sky Patriot ("covering issues of government corruption . . . the militia movement, the Freemen"). According to the posting, whose writer was identified only as Monte, no "media blackout in the . . . case has been more complete than the blackout imposed over the name of Daniel Spiegelman; it is as if he has stepped off the face of the planet. Who is Daniel Spiegelman? What connection does he have with Oklahoma City? Is he John Doe Number Two?"

When I posed some of these questions to Jean Ashton during a phone interview in early 1999, she said she honestly didn't know—and what's more, she didn't particularly care. What mattered to her was, first, that Spiegelman had been extradited to the United States, where in April 1997 he pleaded guilty to a variety of charges at the U.S. District Court in Manhattan; second, that Judge Lewis A. Kaplan had set aside federal sentencing guidelines and handed out a harsher-than-normal prison sentence of five years, arguing that the crimes were more serious than "a simple theft of money"; and, third, that law enforcement agents had recovered much of the plunder.

All of that was good news, but when I asked her about the fate of the stolen maps, she turned glum. Of more than 250 maps Spiegelman was thought to have stolen, fewer than 50 had been recovered, she said. Worse still, many of those had come from an extremely rare copy of the Dutch cartographer Joan Blaeu's *Atlas Major* of 1667, valued, in its former state, at a minimum of three hundred thousand dollars. Only two intact German-language editions of the book had existed anywhere in the world, she explained, and Columbia's copy was all the more unique because it had once belonged to Alexander Anderson, a prominent early-nineteenth-century engraver who supplemented the original with his own maps. "In many ways," she said, "the Blaeu was the most serious part of the whole theft."

*T*HERE ONCE WAS A MAN WHO THOUGHT CLASSIFYING A criminal was as easy as assigning a Dewey decimal number to a book. His name was Johnny Jenkins, and he was famous in rare books circles, both as a flamboyant dealer from Texas and as president of the Antiquarian Booksellers Association of America. In 1982, when serving as head of security for that organization, Jenkins wrote an article claiming that all library thieves could be categorized into one of five "basic types." There was the Kleptomaniac, the Thief Who Steals for Himself, the Thief Who Steals in Anger, the Casual Thief, and the Thief Who Steals for Profit. It was simple to get inside the head of a crook, Johnny Jenkins declared, because criminals "tend to exhibit classifiable characteristics."

Six years after he wrote those words, this same Johnny Jenkins was found floating in the Colorado River, dead from an apparent suicide. Not long before that, news had broken that a number of the very rare and expensive documents Jenkins had been selling over the years were forgeries. Many mysteries surrounded the case—not the least of which were whether Jenkins was personally involved in creating the fakes, and, if so, why a man of his stature would have felt compelled to do such a thing. They remain mysteries still. In both his controversial career and his enigmatic demise, Johnny Jenkins would defy simple classification. "I suppose he was a man who lived two lives," one of his old friends told Calvin Trillin of *The New Yorker.* "Or maybe three."

I keep this tale in mind whenever I'm tempted to make generalizations about the motivations of modern-day map thieves. While at first glance a few of them do seem to fit nicely into those criminal categories Johnny Jenkins invented, on closer inspection most appear to have been driven by complex and contradictory impulses. (A New Hampshire jury, for instance, rejected William McCallum's insanity defense, determining that while he may have had a preference for stolen underwear and hot maps, he was not a kleptomaniac.) Still, if I had to pick out a single overriding motive for map theft, I would opt for the one that's been driving cartographic crime for centuries: simple greed. As a sixteenth-century Spanish diplomat wrote about Lopo Homem, official Portuguese Master of Sea Charts and unofficial smuggler: "[He and his assistant] have orders to make no chart for anybody but the King. But sometimes they venture at a price."

Yet there's another motivation, one that Johnny Jenkins failed to mention but that may have been a factor in his own fall from grace. Simply put: the chance of getting caught in a book-related crime is fairly small, and the chance of getting serious punishment is still smaller. This is especially true of map theft—in part because of the very structure of the trade. The

Columbia librarian Jean Ashton—who is not particularly hopeful that her missing maps will ever be returned—explained the situation like this: "The market for maps is—what shall I say?—much more *uncontrolled* than the market for medieval manuscripts, for example. These early printed maps are bought not just by scholars or the rarefied collector. They're sold as decorative objects, not as scholarly objects. And they may be sold in all kinds of ways that fall outside the system of security by the dealers and collectors of other kinds of rare materials—at flea markets, for example. So it's far less likely that people can track down an individual map."

Ashton stopped short of directly blaming dealers, aware that even the most scrupulous ones often find it impossible to recognize a stolen map once it has been cut out of a book. But when I discussed the issue with Graham Arader, the longtime alpha male of the map kingdom was not so diplomatic. He insisted that his competitors are often less than thorough about checking the provenance of their merchandise. "It's simple," he said. "All you say is, 'Where did you get this map?' Then you listen to the story and you say, 'Do you mind if I check your sources?' And then if he starts waffling, you say, 'Sir, get the hell out of my gallery!' And if you really think he stinks, then you turn him in."

On a number of occasions, Arader has practiced exactly what he preached. It was he, for example, who sparked the investigation into Skeet Willingham, notifying University of Georgia authorities in 1985 that someone had offered him a letter by the Revolutionary War hero Nathanael Greene—a manuscript ostensibly in the university's collection at the time. Still, Arader had to concede that even he has made mistakes, such as when he bought sixty-two maps from Andrew Antippas, unaware that more than forty of them had been stolen from his own alma mater. The fact that they were hot did not come to light until Arader offered some of them for sale to a Texas collector named Walter Reuben, who recognized the maps as belonging to Yale. "Antippas got me," Arader conceded. "He definitely tricked me."

But, in his usually pugnacious style, the map mogul laid much of the blame for the ongoing wave of thefts squarely on the shoulders of librarians, who he claimed are simply not vigilant enough. "Most librarians are incompetent, boring, and dull," he said. "And they have this easy life. Many of them view their collections as their personal fiefdoms. But really, they don't look after their material. You know, it's not hard to tell the difference between a thief and somebody who's legitimate. If you're not intelligent enough to see these guys coming, then you shouldn't be a curator."

It's easy to see why librarians don't care much for Arader. Even so, few of them would deny that rare books rooms face serious security problems. And while the curators—especially those who think they are somehow above the nuts-and-bolts work of preventing theft—certainly deserve a good part of the blame, so do the institutions themselves. Well into the 1990s, for example, many university libraries lacked surveillance cameras for even their most rare and valuable collections, a shortcoming school administrators largely chose to ignore. Protecting books is "simply not what we value as a society," observed the security expert Katharine Leab. "If there is a choice between providing funds to inventory the stacks or to get a new bus for the football team, what do we do? Well, we get a new bus for the football team."

But if the map industry and map libraries have not done enough to discourage theft, neither has the legal system. Willingham served only thirty months of a fifteen-year sentence before getting out on parole. Still, his punishment was comparatively harsh. McCallum, who could have served the rest of his life in prison for the nearly seventy counts against him, got a three-to-six-year term. Fathers Huback and Chapo each received only eighteen months. Antippas got one year, while Opie got just six months for his decade-long crime spree. Glaser, a habitual offender, received only three years of *probation* for his 1992 conviction. And even Spiegelman—whom the judge went out of his way to punish—got only five years in prison (of which he had already served more than half) plus a $314,150 fine (of which he may never get around to paying more than a small fraction).

"If you steal a Picasso or Rembrandt or any piece out of a museum," observed Carol Miller, the former head of security for the Free Library of Philadelphia, "there is going to be publicity and there is going to be serious time given. But if you steal things out of a book, the sentences are just ludicrously low. People say, 'Well, it's just a book.' What kind of message is this sending to people who are thinking about doing this? We're saying, 'Go ahead, because you're not going to get in too much trouble.' It's sort of a joke."

As the prices of rare maps soared in the mid-1990s, the security problems that had plagued libraries and the rare books trade for decades remained largely unsolved or unaddressed. Which meant that the stage was set for the most prolific cartographic criminal in American history to make his quiet entrance.

GILBERT BLAND STOLE IT FROM THE UNIVERSITY OF Washington. John Ogilby stole it from Arnold Montanus, who stole it from somebody else. Ogilby's 1671 map of Virginia— one of those that Bland allegedly pilfered in Seattle—offers proof that the cartographers of yore were themselves shameless crooks. As the biographer Katherine S. Van Eerde so delicately put it, Ogilby "failed to mention" that he had borrowed the map from Montanus, who, in turn, forgot to announce that he had borrowed it from earlier cartographers. Save for a few decorative details, the Ogilby and Montanus maps of Virginia are dead ringers for a 1630 work by Willem Janszoon Blaeu, the leading Dutch cartographer of the seventeenth century (and father of the Joan Blaeu mentioned earlier in this chapter). For his part, Willem Blaeu tried to put a stop to such plagiarism. According to Lloyd A. Brown, Blaeu "made a special plea to . . . Holland and Friesland for protection against the vultures who were pirating his maps, asserting that he could support his family by honest means, with God's mercy, if certain persons would desist from copying all his newest maps before the ink on them was dry." But Blaeu himself was not above theft, as his own 1630 map of Virginia makes clear. He bought the plate for the map from the rival Hondius firm, replacing the Hondius name with his own (a change that left faint traces of the erased letters on his version). Yet the Hondius family had no rightful claim to the work either. Of all those who took credit for the map, the only legitimate one was the English adventurer John Smith, who drafted it in 1612.

Smith had a big advantage over his imitators: he had actually been to Virginia. But what differentiated the creators from the copiers often had nothing to do with firsthand familiarity with the terrain. Zebulon Montgomery Pike, for instance, made extensive explorations of the American West, whereas Alexander von Humboldt never set foot there. But Pike's 1810 work, *A Map of the Internal Provinces of New Spain,* presented as being based on "sketches of . . . Captain Zebulon M. Pike," was in fact plagiarized directly from a map that Humboldt had left at the State Department during a visit to Washington in 1804.

In an 1813 letter of apology to Humboldt, Thomas Jefferson regretted that Pike "made ungenerous use" of the map, insisting it had been done "on a principle of enlarging knolege [*sic*] and not for filthy shillings and pence." But in fact a hunger for "enlarging knowledge" was precisely what separated the two mapmakers. Pike, a career soldier in the U.S. Army, was good at following orders but lacked a necessary zeal for getting at the crux of things. During his 1806–7 expedition, he set out with much hoopla to climb a "Grand Peak" he had spotted in the distance. But after experiencing a bit of cold and hunger, he abandoned the project, miles short of the summit of what is now known as Pikes Peak. Humboldt, by contrast, was bent on penetrating the unknown. When faced with a daunting climb of his own—Ecuador's Mount Chimborazo—he braved rough weather to reach "the highest altitude ever attained by any human being in history until that time," according to his biographer Douglas Botting. He was not only a great explorer—covering more than six thousand miles of South America during one five-year period—but also a brilliant physiologist, zoologist, botanist, anthropologist, archaeologist, and meteorologist. Today he is perhaps best remembered as a founding figure in modern geography, a subject that fired his imagination even as a boy. "The study of maps and the perusal of travel books," he once wrote, "aroused a secret fascination that was at times almost irresistible, and seemed to bring me into close relationship with distant places and things."

It was this same inquisitiveness that enabled Humboldt to produce a map of the West that the expert Carl I. Wheat described as "a truly magnificent cartographic achievement." Humboldt could map a place he had never seen—using, in his own words, "a great variety of data" instead of "vague suppositions"—because he was blessed with the sort of intense curiosity that most of us experience so infrequently it often seems to come as a surprise. I'm not talking about the kind of curiosity that *invites* but about the kind that *demands*, not about the kind that says *I wonder* but the kind that says *I must know.* The kind that makes you immerse yourself in a subject, ponder it over and over until you are able to make sense of it for others and, in so doing, give your own life new meaning in some small way. Under such a spell, humans can accomplish the extraordinary. One need not be a genius like Alexander von Humboldt to make unseen worlds appear.

THE MAN WHO BROUGHT GILBERT BLAND TO JUSTICE was neither a high-profile FBI agent nor a flashy detective on a big-city police force. He was a campus cop, not too far up on the law enforcement food chain from the lowly private security guard. Nonetheless, Thomas W. Durrer had a keen mind and, more important, a roiling sense of curiosity. As an investigator for the University of Virginia Police Department, Durrer had little in common with the shoot-'em-up stars of tabloid TV cop shows; his work involved mostly mundane matters such as dormitory break-ins or stolen backpacks. But he took pleasure in the detail work that others might find boring, tracking down all the scattered little pieces of evidence, then putting seemingly unrelated facts together until they told a complete story. And, every now and then, something big would happen in Charlottesville and Durrer would luck into the kind of case where, as he put it, "you just have to kind of use your imagination and see how far you can go with it."

In early December 1995 he landed just such an assignment. Officials at the Alderman Library reported that at least seven rare maps were missing, including eighteenth-century works by the cartographers Herman Moll and Andrew Ellicott. Librarians had discovered the loss after hearing from Johns Hopkins officials, who had noticed references to the Alderman in the map thief's notebook. A check of the records confirmed that someone using the name James Perry had indeed paid visits to the Alderman on December 5 and 6, right before Gilbert Bland's ill-fated stop in Baltimore.

The case seemed tangled and slightly bizarre—everything Durrer could have wanted. "I knew this was probably going to be broader than some of the other cases I had worked," he said. "I knew it was going to be in-depth. And that type of investigation is, for me, fun to work. It's what I love to do."

Solid in stature, with a touch of both Southern gentleman formality and good-ol'-boy waggishness, the fifty-two-year-old Durrer has the air of a shrewd small-town sheriff. And in this investigation, at least, his hunches seem to have been dead-on from the start. Because the thief had apparently visited the library on two consecutive days, Durrer correctly concluded that he had spent the night in town. "I just started checking local motels and found out where Bland had stayed," he said. "It was a Howard Johnson's in Charlottesville, and he paid for his room by charge card. I tracked his address through his charge card and found out where he lived. I then confirmed his address and obtained a picture of him through the Florida Department of Motor Vehicles. People up here at the library made positive IDs on his picture."

Durrer quickly began to share his findings with other universities. "I don't think any of us knew how broad this would get before it was over," he said. "At the time, I thought maybe I had a collector who stole a few maps. Once I got into digging into him, I saw a pattern of him moving from one university to the next. It was obvious what he was doing."

Durrer said that some institutions, such as the University of Florida, were completely unaware that they had been robbed until he alerted them to the possibility. Such discoveries spurred him on. "As this case broadened out, it just made me go after it a little quicker, a little harder. As I found out one thing, two or three more things would show up."

His perseverance paid off. With the list of alleged victims growing nationwide, the FBI entered the investigation, armed with hard evidence linking Gilbert Bland to a crime scene. You could argue that collecting this evidence had been a matter of simple, solid police work, that any investigator from the dozens of places Bland had visited could have done the same thing. Yet the fact remains that only one man took the initiative. Thanks to Thomas Durrer's curiosity, the map thief would eventually wind up in jail.

FEW MONTHS AFTER MY VISIT WITH VERA BENSON AT the American Map Corporation, I called her with a tale to tell. I had just been looking at two early-eighteenth-century maps showing most of what is now the United States. The first one, by the cartographer Guillaume Delisle, was based on firsthand information gathered by French explorers. The second one was—with the exception of a priest, a pelican, a picture of Niagara Falls, a family of Indians, a couple of buffalo, and a pair of naked buttocks—a direct rip-off of the Delisle work. The man who made it was J. B. Homann, that foppish old cartographer whose portrait was perched behind Benson's desk.

When I informed her that she was harboring a known map thief, Benson began to chuckle. "Well, these things have happened throughout history"—she sighed—"but I'm terribly disappointed that my German countryman would have done such a thing."

As we both laughed, I remembered something Benson had told me during my trip to Queens. She had explained that these days a certain amount of cartographic appropriation is considered not only acceptable but necessary. Her company, like its competitors, copies maps all the time—the so-called base maps from the U.S. Geological Survey, which "everyone uses as sort of a framework or skeleton to build their own maps." With the basic shape of the world no longer a matter of debate, the challenge for cartographers now comes from accurately and stylishly modifying these base maps by eliminating "stuff we don't need, such as elevation lines or other features that are not appropriate for a street map," then adding the latest data from satellite photos and a variety of other sources.

"We are still refining the mapping that exists," Benson had concluded, "but nobody has to make a map by going out and surveying the land anymore."

Recalling those words, I laughed even harder—this time at the absurdity of my own situation. You see, the quest I had launched all those months ago at the coffeehouse—then a matter of simple curiosity—had become an increasingly complicated and frustrating endeavor. I now found myself in the midst of a peculiar cartographic project, one for which there were no base maps. My task was even more of a stretch than Alexander von Humboldt's attempt to chart a place he never saw. I was trying to map the life of a man—an anonymous and elusive man, a man I did not know, and a man who had demonstrated no desire to meet me. And even all that might not have been so bad if I had somehow been able to find a way inside his head, to put myself in his shoes. But Bland and I were very different people. Other than a few shared superficialities—both of us white males, both right-handers, both map lovers—our common frames of reference were few. We came from contrasting family backgrounds (Bland: broken home, me: two stable parents), were of different generations (Bland: turbulent 1960s, me: dull 1970s), had opposite dispositions (Bland: extreme introvert, me: strong extrovert), and battled dissimilar demons (Bland: risk taker, me: cautious to a fault). The more I knew of him, the more he seemed a stranger. *Stranger*—a word linked to notions of geography, its etymological root being "beyond the usual bounds or boundaries." That was Bland: beyond my boundaries. Searching for him had transformed me into a land surveyor of a land unknown and perhaps unknowable, never quite sure what I was doing or where I was going. In my most ludicrous moments I felt a creepy bond with another land surveyor—K., the protagonist of Franz Kafka's *The Castle*—a man *haunted by the feeling that he was losing himself or wandering into a strange country, farther than he had ever wandered before, a country so strange that not even the air had anything in common with his native air, where one might die of strangeness, and yet whose enchantment was such that one could only go on and lose oneself further . . .*

CHAPTER 8

Pathfinding

$\mathcal{S}$ILENCE IS A WILDERNESS. I TRUDGED THROUGH it many times as I tried to map Gilbert Bland's world. It was there after every letter I sent him, there when I wrote his lawyers and when I wrote his wife, there when I followed all those letters with phone calls. It loomed in front of me at every turn, vast and unyielding, making distances difficult to measure, directions hard to gauge, landscape tough to decipher. It was daunting, this wilderness, yet strangely seductive. The more I encountered it, the more I was determined to overcome it. Not that I particularly begrudged Gilbert Bland his silence. In an age when almost everyone is willing to go public about almost everything, I could not help but feel admiration for something so foreign as to seem exotic. Sometimes, as it spread out on every horizon like the sands of the Empty Quarter, obscuring all that lay below, I found it almost breathtaking. Eventually,

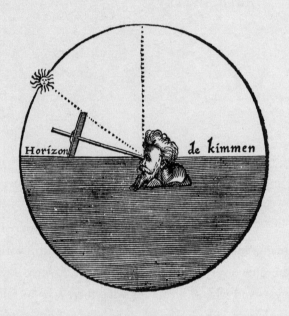

Horizon

de kimmen

A GRAPHIC FROM WILLEM JANSZOON
BLAEU'S *THE LIGHT OF NAVIGATION*
ILLUSTRATING HOW TO FIND YOUR
LATITUDE WITH A TOOL CALLED
A CROSS-STAFF.

though, I had to face the fact that it was not some sort of mirage, that it would not go away, and that if I did not conquer it, it would conquer me.

Curious to tell, it was the map thief himself who inadvertently showed me how I might do it. I had come to believe that when Gilbert Bland cut into all those books, he unloosed not just pieces of paper but stories, hundreds of stories centuries old, about the discoverers and scientists and surveyors and kings and queens and cartographers whose lives were pressed into those maps as firmly and indelibly as the ink, stories that were slowly intertwining in my imagination with the map thief's own tale. And so it was that I began to commune with the dead in search of a guide, someone whose past would light my journey through the shadowy wasteland of the present. Ultimately, I chose a man famous for navigating ground that was, as he once described it, "singularly unfavorable to travel." He was called the Pathfinder.

John Charles Frémont probably did more to popularize the American West than any other explorer of the nineteenth century. Friend and fellow adventurer of Kit Carson, son-in-law of the powerful Senator Thomas Hart Benton, and, later in life, first presidential candidate of the fledgling Republican Party, Frémont earned his legendary nickname by leading several Army Corps of Topographical Engineers expeditions to survey and map little-known parts of the West. He covered more ground than any other government explorer, including Meriwether Lewis and William Clark, and his published accounts of the journeys made him a national celebrity.

Historians love to point out that the Pathfinder did not, in fact, find many paths. The Pulitzer Prize winner Bernard DeVoto, for example, insisted that Frémont "did little of importance beyond determining the latitude and longitude of many sites which mountain men [already] knew." (Not to mention the Native Americans.) True enough—but this view ignores an important point. The Pathfinder understood that to truly conquer a wilderness you must not only survive in it yourself but make sense of it for others. His maps—usually drafted with or by the gifted cartographer Charles Preuss, who accompanied him on a number of missions—helped fuel one of the greatest migrations in American history. Most notable among them was the Preuss-Frémont map published with the 1845 *Report of the Exploring Expedition to the Rocky Mountains*. Presenting for the first time a coherent picture of the continent from Kansas City to the Pacific Coast, it took on vital importance for those who followed Frémont to the region—the builders of railroads, the settlers of land, the seekers of gold. "The map . . . radically and permanently altered western cartography," wrote Carl I. Wheat in his *Mapping the Transmississippi West*. "There would appear a few throwbacks, of course, but with thousands of copies of Frémont's authoritative report and map in print, no cartographer could find reasonable excuse for not fairly representing the main features of the West."

Frémont's philosophy of mapmaking was simple. He once offered the following instruction on "fixing on a small sheet the results of laborious travel over waste regions, and giving to them an enduring place on the world's surface:

First, the foundations must be laid in observations made
in the field; then the reduction of these observations to
latitude and longitude; afterward the projection of the
map, and the laying down upon it of positions fixed by
the observations; then the tracing from the sketch-
books of the lines of the rivers, the forms of the lakes,
the contours of the hills.

I wondered if this straightforward formula for charting unfamiliar land would also work for charting an unfamiliar life. By
this time I knew exactly what Frémont had meant by "laborious
travel over waste regions." I had accumulated stacks of newspaper and magazine articles about the case, interviewed scores
of people who had come in contact with Bland, retraced the
path of much of his crime spree, and brought together hundreds of pages of public documents relating to his personal, financial, and criminal history. At long last my fieldwork was
almost complete. It was time to begin "laying down . . . positions fixed by the observations."

Here, too, I took inspiration from the Pathfinder. Frémont
and Preuss were not afraid to leave huge white areas on their
maps. They understood that no single map can say everything
about a place, so they cautiously chose to incorporate only that
information which came from their own observations. Still,
they made some mistakes. On one map they combined the
Great Salt Lake and Utah Lake as a single body of water. On another they conveyed the false impression that the interior basin
of present-day Nevada was encircled by mountain ranges. They

knew that after all the measurements are complete, you still have to make some judgment calls. The Pathfinder inspired me to trust my own instincts in identifying the landmarks of Bland's life—and to let the blank spots stand.

So I set about the task of drafting my map, beginning with two coordinates close in time but half a world apart.

Latitude: 40.40 N. Longitude: 74.06 W. June 17, 1968

At 10:55 P.M., police in Bayonne, New Jersey, arrested a skinny, redheaded kid from nearby Ridgefield Park for possession of a stolen motor vehicle. He was eighteen years old but looked two or three years younger. In his now-faded mug shot, he sports a white T-shirt, a tight-lipped scowl, and a conservative mop top, half Beatles, half *My Three Sons*. His eyes are filled less with fear or anger than with numbness, as if he is looking through the camera instead of at it, as if he is trying to daydream himself right out of the police station.

He was in serious trouble. In 1968, conviction on the charge of possession of a stolen motor vehicle in New Jersey carried a maximum sentence of a five-thousand-dollar fine and ten years in prison—not a happy prospect for a kid who had just graduated from high school. Then again, the young Gilbert Bland was apparently no stranger to tough times. He had been born in 1949 in Indianapolis, where his father, Gilbert Lee Joseph Bland (listed on the boy's birth certificate as a baker), and his mother, a nineteen-year-old New Jersey native named Julia Patricia Bland, lived in a tiny, one-story house in the shadow of an

International Harvester foundry. Today that house sits in a bleak working-class neighborhood, around the corner from Ron & Deb's Tobacco Outlet and the Paradise Cove, a self-described "gentleman's club." Whatever happiness his parents found in that place did not last long. The couple separated when Bland was only three years old, according to Bland's lawyers in a later case. Some two years later, in 1954, Bland's father sued his mother for divorce. After Julia Bland failed to show up in court on three separate occasions, a Marion Superior Court judge ruled that "the allegations of the complaint alleging statutory grounds for a divorce are true." Surviving court records do not make the specific nature of those allegations clear. But whatever the case, Julia Bland, who was granted custody of the boy, took him back to her home state of New Jersey and at some point remarried.

By the time of his arrest, the teenager lived in a sturdy house in Ridgefield Park, a New York suburb on the banks of the Hackensack River. It was a pretty town with tree-lined streets, quaint slate sidewalks, and one of the longest-running continuous Fourth of July parades in the country. Bland's life there, however, appears to have been less than all-American. His mother's second husband "was both verbally and physically abusive to him," according to Bland's lawyers in a later case. No matter how bad his home life was, though, it could not have seemed worse to him than the prospect of prison. I don't imagine he slept well in the police lockup that night.

But in Bayonne Municipal Court the next day, Bland got a lucky break—the first of many he would receive in his dealings with the law. Prosecutors amended the charge to the lesser of-fense of unlawful taking of a motor vehicle. He pleaded guilty and was given a hundred-dollar fine. After reclaiming his per-sonal effects—$1.25 in cash, keys, a wristwatch, and a comb—Gilbert Bland emerged into the muggy Meadowlands air, a free man.

How had he managed to get off with such a light sentence? In plotting my map, this is one of those places where I have trouble making sense of the terrain. Public records do not indi-cate why the charge was reduced, and when I tracked down re-tired judge Harvey L. Birne, who was on the bench that day, he had no recollection of either the case or the defendant. "It's more than thirty years ago," sighed Birne, who is now in private legal practice.

He was, however, willing to venture a guess about what hap-pened: "If there wasn't a good proof of intent to steal—if it was just joyriding or taking of a friend's car—[prosecutors] would downgrade it. If it's a downright theft, to my experience, they wouldn't downgrade it. . . . The sentence seemed quite low. I would imagine it was a minor case and first offense."

That makes sense. Yet I can't help but wonder whether a much different scenario played out in court that day—one that would have had a lasting impact on Bland's life. My suspicions are based on a single tantalizing fact: just over a week after Gilbert Bland walked out of jail in Bayonne, he walked into the base at Fort Dix, New Jersey, for induction into the U.S. Army. These two incidents may be unrelated—but then again they may not. New Jersey law enforcement sources told me that, during the Vietnam era, local prosecutors and judges were sometimes known to offer young defendants a deal. It went roughly like this: Son, you're going to serve time no matter what. You can serve it either in jail, wasting away your youth, or in the military, defending your country. It's your choice: join the U.S. Army or join the New Jersey penal system.

Was Bland offered such a deal? Apparently so, according to one source who knew him in the past. The source, who spoke on the condition of anonymity, said Bland himself had always insisted that the threat of jail was what drove him to the Army. Further—though hardly conclusive—support for this version of events comes from government documents. Although privacy laws prevent the military from disclosing whether a soldier signed up or was drafted, Bland's attorneys in a later case maintained their client had "enlisted in the United States Army." Bland's Army serial number, obtained from law enforcement records, also indicates that he was an enlistee. Yet the notion of a gung ho Gilbert Bland joining the military out of patriotism alone strikes me as out of character. He never seemed to be much of a joiner in anything, much less in the horror of Vietnam, and his political views were decidedly left-leaning, according to people who knew him later. It seems more likely that he was coerced to enlist, whether for economic, familial, or legal reasons.

For his part, Judge Birne claimed no memory of a deal with Bland (or anyone else) in which enlistment was offered as an alternative to prison—"not that I object to the proposition." If there was any special treatment, he said, it would have been because Bland was already Army-bound when he appeared in court. "If I knew he was going in [to the military] definitely," said Birne, "then perhaps we would have eased the pain and let him go."

But regardless of whether it led directly to his tour in the Army, the episode appears to have been a critical one for Bland. Arrested for the first time, he had learned an important lesson about how the judicial system could be manipulated to his advantage. He would not forget it. But there was another big lesson he had not learned. He had not been dissuaded from stealing things.

*M*Y FINGER IS RUNNING OVER ONE OF THOSE THINGS (or at least a reproduction of it) right now. It is titled *Map to illustrate an exploration of the country, lying between the Missouri River and the Rocky Mountains, on the line of the Nebraska or Platte River. By Lieut. J. C. Frémont, of the Corps of Topographical Engineers.* The Pathfinder drafted it, with the help of Preuss, to illustrate his report on the journey of 1842, his first great expedition of the West. My fingertip is lightly retracing the course of that trip, beginning at the point where the North Fork and South Fork of the Platte River break apart in what is now Nebraska, then gliding west to Fort Laramie in present-day Wyoming, and continuing on to the South Pass, the Oregon Trail's principal route through the Rockies. From there I run my finger north into the Wind River Range, drawn in dramatic hachure-style relief. This was where Frémont concluded the westward leg of his journey with typical bravura, scaling what he had determined to be the "loftiest peak of the Rocky Mountains" and planting a specially designed American flag on its summit. The mountain, it turned out, was not even close to being the tallest, but it was the symbolism that mattered. The climb became a rallying point for Manifest Destiny and helped make the Pathfinder's a household name.

Frémont's heroics, however, are not the reason I am now leaning over this map. Just the opposite, in fact. As I move my finger south, then east, following the expedition's return route, I am looking for a place where a dark and self-destructive side of the Pathfinder's personality emerged, a place that closely corresponds with the one I am trying to plot on my map of Gilbert Bland. Yes, here it is now—right where the Sweetwater River and North Fork of the Platte converge, a point southwest of present-day Casper, Wyoming. Don't bother looking for this spot on a modern map, because it's no longer there. In its place is something called the Pathfinder Reservoir—an odd double homage to John Charles Frémont, both memorializing his accomplishments and literally blotting out the site of one of his greatest blunders. This was where Frémont did something "stupid" and "foolhardy" (in the words of Preuss), something that risked not only the success of his expedition but his entire future.

You see, that was another reason I chose the Pathfinder as my guide: he reminded me a lot of Bland. Perhaps it was a matter not so much of common personality traits but of a shared emptiness—an emotional vacuum that both men seem to have spent their lives trying to fill. Like Bland, Frémont came from a broken home, in which his father was taken from him at a young age. The Pathfinder was the illegitimate child of a married Virginia society woman and her lover, the mysterious Charles Fremon, who, according to the varying accounts, taught French, gave lessons in swordsmanship, was a "dancing master," and/or did "occasional upholstering." Fremon ran off with the boy's mother—leaving her shamed and penniless—then died when John Charles was only five years old. Soon after, this young boy, like Bland, began dabbling in imaginary creatures. Early in life he re-created himself by adding a *t* to the end of his name, as well as an accent on the *e*. "No one ever knew why," wrote John Moring in *Men with Sand: Great Explorers of the North American West.*

But Andrew Rolle, author of *John Charles Frémont: Character as Destiny,* was willing to offer some clues. In his 1991 biography Rolle used "psychiatric techniques . . . to seek a better understanding of the man." He attributed much of the Pathfinder's adult behavior to a "fragmented sense of selfhood" stemming from the separation from his father. "Because Frémont's formative years of childhood were obscured by long shadows," concluded Rolle, "he fashioned an image of himself based in large measure upon fantasy."

That image, of course, was of a flamboyant adventurer. But in private the Pathfinder had a strikingly different demeanor. "In fact," wrote Rolle, "even while on his expeditions, some of Frémont's contemporaries used the terms *bland* and *gentlemanly* as well as *quiet* and *retiring* about him." The italics are Rolle's—

but the words sound like those used time and again to describe the map thief. And, like Bland, Frémont was a loner with a decidedly antiauthoritarian streak that often led him to break rules.

Frémont and Bland, however, were polar opposites in at least one important respect: the Pathfinder's life was as conspicuous as the map thief was silent and anonymous. But this, too, was to my advantage. It occurred to me that, like the best wilderness guides, Frémont might serve as my interpreter. I began studying his life, hoping that it would give me perspective on Bland's. This process, which took weeks, has led me to where I am now—at my desk, listening, once again, to the strange song of an old map, my finger tapping down on a spot that no longer exists.

It was there, on August 24, 1842, in what the biographer Allan Nevins called a moment of "reckless impetuosity," that Frémont decided to load the mission's precious records and scientific equipment into an experimental rubber boat and, without any advance scouting, run the swollen rapids of the Platte. There was only the flimsiest scientific justification for the effort; the Pathfinder seems to have been mostly interested in seeking thrills. He found them. Pushing off from shore with five other men—three of whom could not swim—he was soon battered out of control downstream. At one point Charles Preuss waded ashore with the expedition's chronometer and tried to walk with the precious longitude-measuring instrument to safety. Unable to traverse the rocks, the mapmaker had to climb back in the boat, which was now in a more perilous position than ever. "To go back," Frémont wrote in his report on the expedition, "was impossible; before us, the cataract was a sheet of foam; and shut up in the chasm by the rocks, which, in some places, seemed almost to meet overhead, the roar of the water was deafening. . . . We cleared rock after rock, and shot fall after fall, our little boat seeming to play with the cataract."

One more rock loomed ahead, hidden from view beneath the foam. This was the inevitable one, the one with the Pathfinder's name on it, the one that would flip the boat, nearly killing Frémont and his men and causing them to lose their sextant, their large telescope, two of their compasses, most of their food and clothing, and a journal containing important weather and cartographic data. But it could have done worse harm, that last rock. It spared not only his life but most of the scientific records. Had those observations disappeared beneath the rushing waters, Frémont's career might have sunk with them. Yet

DETAIL FROM THE TITLE PAGE OF A 1617 ATLAS BY PIETER VAN DEN KEERE.

as that underwater boulder drew near, the Pathfinder was not thinking about any possible disasters. When he recalled the event later, he remembered only the glorious feeling of being "flushed with success, and familiar with the danger."

I suspect some similar emotion was rushing through Gilbert Bland's mind when, according to the FBI, he crept out of the University of Rochester library, carrying a stolen copy of this same Frémont-Preuss map. The alleged incident took place on October 10, 1994—fairly early in his crime spree. But perhaps, like the Pathfinder on that river 152 years earlier, he already felt there was no going back. And maybe that was exactly what he craved: the rush of being pulled along by an overpowering current, of closing in on catastrophe at each turn, of defying the odds, of feeling intensely alive. *Flushed with success, and familiar with danger*—we've all known this particular joy at one time or another. Some people get it from rock climbing, skydiving, bungee jumping, or any of the other sports that allow us to flirt with our own funerals; others find it in more common pursuits, from speeding on highways or soaring on roller coasters to betting on horses or cheating on spouses. But while most of us require only the occasional taste of peril, others are addicted to it. The Pathfinder, for example, seemed to learn nothing from his near-disaster on the Platte. He sought out similar situations time and time again, both in his professional life (where, in 1848, his insistence on trying to cross the Rockies in midwinter left ten men dead and forced others to resort to cannibalism) and in his personal life (where his land speculation deals and other risky ventures left him penniless at his death in 1890). His biographer Andrew Rolle once again traced this behavior back to the loss of his father in boyhood:

> Bereft of two nurturing parents who could understand his needs, Frémont repeatedly, as we have seen, acted as though he wanted to validate his illegitimacy and lack of a complete family during childhood. . . . Child psychoanalysts, after studying hundreds of patients, have established the presence of unresolved bereavement stemming from loss of a parent in childhood. These modern clinicians have found that serious emotional impairment occurs throughout the lives of such patients.
>
> Recent findings regarding early separation and loss are thus crucial to understanding that later seemingly inexplicable conduct which often expresses a disguised repetition of early sadness. In such cases outbursts against authority are frequent. Also, persons who repeatedly place themselves in dangerous circumstances may be attempting to work their way back mentally to an unresolved conflict.

While I have no idea whether Rolle was right about the Pathfinder, I tend to be wary of writers who depend too heavily on pop psych generalizations about their subjects. And, since I have absolutely no professional expertise in matters of the mind, I am neither willing nor able to make any such sweeping statements about Bland. All I can safely conclude is that he, too, possessed a self-destructive attraction to danger, and that it had been with him for a long time. Perhaps it was already there on that summer night in Bayonne in 1968. If so, it would have been one more thing he had in common with John Charles Fré-mont—who also got in trouble at age eighteen, when he was dismissed from college for "habitual irregularity and incorrigi-ble negligence." Both Frémont and Bland would then turn to the military—in which both would use their keen intelligence to earn technology-focused assignments, and which both would eventually leave under a dark cloud (in Frémont's case, resignation after a court-martial for "mutiny, disobedience, and conduct prejudicial to military discipline"). But, long before that happened, each man would head off to a place that would change his life—a place, as Frémont put it, "of strange scenes and occurrences." For the Pathfinder, of course, that place was the American West. For the map thief, it was a very different sort of wilderness.

Latitude: 13.46 N. Longitude: 109.14 E. May 4, 1969

This is the way Michael Herr began *Dispatches,* his landmark book on the Vietnam War:

> There was a map of Vietnam on the wall of my apart-
> ment in Saigon and some nights, coming back late to
> the city, I'd lie out on my bed and look at it, too tired to
> do anything more than just get my boots off. That map
> was a marvel, especially now that it wasn't real any-

more. For one thing, it was very old. It had been left
there years before by another tenant, probably a Frenchman, since the map had been made in Paris. The paper
had buckled in its frame after years in the wet Saigon
heat, laying a kind of veil over the countries it depicted.
Vietnam was divided into its older territories of Tonkin,
Annam and Cochin China, and to the west past Laos
and Cambodge sat Siam, a kingdom. That's old, I'd tell
visitors, that's a really old map.

If dead ground could come back and haunt you the
way dead people do, they'd have been able to mark my
map CURRENT and burn the ones they'd been using since
'64, but count on it, nothing like that was going to happen. It was late '67 now, even the most detailed maps
didn't reveal much anymore; reading them was like trying to read the faces of the Vietnamese, and that was
like trying to read the wind. We knew . . . that for years
now there had been no country here but the war.

I thought about those words often as I, too, lay in bed, trying to make sense of Vietnam, or at least the Vietnam experience of a certain Private First Class Bland. Like the regions
covered on Herr's map, the area I was trying to plot on mine no
longer really existed. It was a place where solid facts could not
be trusted and normal rules did not apply, where someone
might enter as one person and leave as another person entirely.
How could I hope to understand Bland's life in a country that
was no country but war? In Herr's words, it seemed like trying

to read the wind. Yet I had to try. The Vietnam conflict had a huge impact on all who took part in it—but in Bland's case it loomed especially large. At least that's what his lawyer would argue more than a quarter century later, when he stood up in a federal court and implied that the war had been at the root of Bland's string of map thefts. His attorney Paul R. Thomson, Jr., claimed that his client "has a pattern of problems . . . largely triggered by depression, a very common problem with post-traumatic stress syndrome." Not that Thomson was willing to supply me with any specifics. As usual, I would be navigating this strange landscape by dead reckoning.

HE MARINES WANT A FEW GOOD MEN. THE ARMY SIG-nal Corps wants a few smart ones. The Signal Corps is responsible for military communications, a demanding discipline as old as warfare itself. Ancient armies used drums, bells, banners, and flags to transmit intelligence, while later militaries employed lights, pistols, pyrotechnics, and homing pigeons, then the telegraph, teletype, radio, radar, and telephone. By Vietnam the Army Signal Corps was "actually ahead of the commercial phone companies at the time. They were the first to use communications satellites," John D. Bergen, an expert on the Signal Corps, told me.

To operate this state-of-the-art system, the Army needed to seek out exceptionally bright soldiers. It found them among incoming troops who were given tests to determine their intelligence and technical aptitude. "It was the equivalent of the college boards for the military. And the higher scores would get into this field," explained Bergen, who wrote *Military Communications: A Test for Technology*, a history of the corps in Vietnam.

Gilbert Bland apparently scored well enough on those tests to be offered the option of joining the Signal Corps and receiving advanced instruction after basic training. He accepted—perhaps because the offer was accompanied by an additional enticement. Joseph Zottarelli, Jr., who served in the same company as Bland, remembered getting "a lot of bullshit from the enrollment sergeants. You know, 'Hey, take these [tests] and you won't be going to Vietnam. You'll probably be going to Germany.' " But once Zottarelli arrived at Fort Monmouth, New Jersey—the same place Bland received his electronics training—he quickly discovered that "most of all the equipment they were teaching us how to use was in Vietnam. . . . I think about 90 percent of the guys that graduated from signal school went over there."

Bland was among them. His military records show that he arrived in Vietnam on May 4, 1969, for a one-year tour. He was a few months shy of his twentieth birthday. "They took guys over in commercial airlines," remembered Zottarelli, who arrived several months after Bland. "And when you got off the plane, they said, 'Hurry up and get off because the plane has to get off the ground.' They didn't want the plane to stay on the ground too long because it was taking mortar and rocket fire. So you got in country and you wondered, What the hell is going on?"

Bland could not have been happy to be there. But he must

have known that most soldiers in Vietnam had far worse jobs. Assigned to the 361st Signal Battalion, he had as his official mailing address Qui Nhon, a once-small fishing town that U.S. forces had transformed into a bustling port city, complete with a naval station and military base—a burg of such wartime decadence that the mayor reportedly turned his official residence into a massage parlor for American soldiers. It's unclear, however, whether Bland actually lived in Qui Nhon, or for how long. His records hint that he may have been reassigned to at least one other locale. This would not have been unusual for members of the 361st, who were spread out in small outposts, which provided communications links around the country. These "fixed station" sites usually consisted of two single-story prefabricated buildings filled with electronics equipment and surrounded by several large billboard or horn-shaped antennas. Some sites were located on the edges of huge military installations, others in more isolated areas. Either way it was not known as a particularly dangerous assignment.

"I really didn't come under fire that often," said Frank Dapuzzo, who served in the same company as Bland. "Occasionally, [the enemy] would lob in some rockets and some mortars—but believe me, there were a lot worse conditions than what I had to go through. Some of the people who really had to go through bad times there, they would look at us and say we were on the gravy train."

Dapuzzo, like several other former members of the company with whom I spoke, has no memory of Bland—which may have less to do with the map thief's inconspicuousness than with the unit being so spread out. Dapuzzo, for instance, was stationed at a Korean army base in Ninh Hoa and had little face-to-face contact with other members of the 361st. It's a safe bet, however, that Bland's day-in, day-out life in Vietnam did not differ much from that of the other members of the company. He worked as a fixed-station controller, a twelve-hour-on, twelve-hour-off assignment spent inside an air-conditioned building, in which the main enemy was likely to be boredom, not the Vietcong or North Vietnamese.

"The environment in the fixed stations in Vietnam by '68 or '69 was very civilized, at least on a base," Bergen told me. "The big fixed stations where these controllers worked were in most cases attached to bases that had sidewalks, PXs, and swimming pools. It would be just like being on a base in the United States. Some of the fixed stations—on places like Monkey Mountain outside of Da Nang—would be more isolated. But even those places, though they wouldn't have a PX or a swimming pool, would still have an environment with air-conditioned dormitories, safe water, generators, and all that kind of stuff. . . . The degree of danger in a job like that compared to an infantryman or a radio-telephone operator with an infantry unit was like night and day."

Gerald Streett, another former member of Bland's company, offered a similar perspective. "You got up; you ran your checks on your equipment; you did maintenance, things like cleaning up outside of the van; you did a daily calibration of your monitoring equipment. And that's about it. . . . If you had to go to a combat zone, this was kind of a nice job to have. I can't say they were fond memories, but there were one hell of a lot of worse places to be."

In fact, said Streett, the enemy actually went out of its way not to fire on his communications site on the edge of an air base, Nha Trang—and for good reason. "We had these sixty-foot billboard antennas. They had red anticollision lights on them, and [the enemy] used those lights as a reference for firing on anyone else in that geographic area. We were at one end of the airfield—and, of course, Charlie knew enough about that base to say, 'Okay, those lights are the 361st; if I go a quarter mile to the right I'm going to hit so-and-so; if I go an eighth of a mile to the left, I'm going to hit someone else.' It wasn't that they loved us, it was just that they thought, 'Hey, these antennas are great!' I mean, they were on twenty-four hours a day. So we were Charlie's best friend."

Nothing in Bland's military record indicates that he saw any direct combat. His medals, for example, were all perfunctory, most of them indicating only that he passed certain skill tests. It seems unlikely that his war experience was the sort that would traumatize a soldier for the rest of his life.

Unlikely, yes, but not impossible, according to Don R. Catherall, a psychologist who serves as executive director of the Phoenix Institute in Chicago, a clinic specializing in treating victims of post-traumatic stress disorder. "There have been cases in which it has been accepted that people in Vietnam suffered traumatic stress even though they were not exposed to a discrete trauma, such as being shot or being shot at, or being in a particular combat action," explained Catherall, himself a

Vietnam vet. "The state of the entire countryside was such that it would be possible for a person to go through a noncombat tour there and still never feel safe. There was a lot of guerrilla activity that couldn't be predicted, and of course there was terrorism going on. Even in the rear, where people didn't carry weapons, there was the possibility of rocket attacks. So it is possible that some individuals remained in a state of heightened arousal, similar to what happens when somebody is actually exposed to a trauma, but that their state of arousal was primarily geared toward anticipating a possible trauma happening. . . . The heart of trauma is often a feeling of helplessness."

And that feeling was not unfamiliar to other members of Bland's company. "You're not actively engaging the enemy, but yet you're not in any position to run, either," Gerald Streett said about working at the fixed-station sites.

In his book on the Signal Corps, John D. Bergen noted that life in the sites "held its own dangers and terrors," many of which stemmed from working day after day "in the windowless confines of a building . . . and being unaware of what was happening outside." Reading those words, it struck me how deathly claustrophobic it might have been for somebody trapped in one of those trailers, the war waiting somewhere in the unseen distance like a huge crimson-eyed beast. Fear of the unknown—is this not among the most ancient and awful of human horrors? Indeed, the scholar G. Malcolm Lewis has wondered whether it was the very thing that led to the invention of maps. For early humans, he wrote, mapping may have "served to achieve what in modern behavioral therapy is known as desensitization: lessening fear by the repeated representation of what is feared. Representing supposedly dangerous terrae incognitae in map form as an extension of familiar territory may well have served to lessen fear of the peripheral world."

It plagues us still, this dread of the dark beyond. "In some cases, the reaction is not to what's outside, it's to what you *perceive* to be outside," explained Catherall, author of *Back from the Brink: A Family Guide to Overcoming Traumatic Stress.* So it's at least possible that the war had a particularly strong impact on Bland. And, as his fellow soldier Joseph Zottarelli, Jr., observed, returning from Vietnam was tough on everyone, regardless of actual combat experience. "It kind of screws you up when you come back," he said. "All the time you were over there, you kind of missed out on everything over here. And there were a lot of changes going on in the States. It took a long time to get back into things."

One thing it did not take Gilbert Bland a long time to get back into, however, was trouble with the law. On October 20, 1970—less than six months after he had returned from Vietnam—he was arrested in Woodcliff Lake, New Jersey, for desertion from the Army. "We got a call of a suspicious vehicle in the orchards; in that era we had a lot of orchards," remembered Woodcliff Lake Police Chief Richard Poliey, a patrolman at the time. "I was given that call. I checked it out and observed a white male who was later identified as Gilbert A. Bland."

Poliey remembered Bland as being "kind of disheveled. He didn't fit the profile for the area. I thought he was a drunk driver." Upon discovering that the auto registration the suspect produced did not match his plates, Poliey ran Bland's name through the National Crime Information Center. "I got a hit that he was a deserter from the military. . . . The MPs came and got him, and that was the end of that."

Then as now, the maximum penalty for desertion was death. Such a punishment would have been extremely unlikely, given Bland's circumstances and the political realities of the time. Nonetheless, he was, by all indications, in serious trouble. Instead of being sent back to Fort Riley, Kansas, where he had previously been posted, he was assigned to nearby Fort Dix. There, according to military records, he was put in a special detachment that Daniel Zimmerman, curator of the Fort Dix Museum, described as "an organizational unit . . . where people would have been assigned who were awaiting a court-martial or awaiting discharge or awaiting transfer to some other custody, say the local police. He would not have been confined to prison, but he would have been confined to the barracks."

But that did not stop him from getting arrested again, on January 21, 1971, this time in his hometown of Ridgefield Park. According to police records, the charges—possession of burglary tools, failure to give good account, and trespassing on private property—stemmed from someone "trying all of the apartment doors" in a local building. He was found not guilty on those charges, but police records nonetheless indicate that he was AWOL at the time of his arrest. He was returned to Fort Dix, but some six months later, on July 6, 1971, the FBI arrested him for allegedly being AWOL once again. What happened next is something of a mystery. Bland's military records show that at least part of his final six months in the military was spent in a "pre-trial confinement facility"—but those same records show that no court-martial trial ever took place. Perhaps it was just too much trouble to court-martial a soldier whose military career was almost over (though privacy laws prevent the military from revealing such information). Whatever the case, Gilbert Bland left the armed services for good on August 31, 1971, at the age of twenty-two, apparently having dodged major legal problems for the second time in his short life.

*I*N THE PATHFINDER'S DAY, MANY MAPS OF THE AMERIcan West contained something called the Rio Buenaventura, a huge waterway flowing from the Rocky Mountains to the Pacific. It was the latest incarnation of a centuries-old myth known as the Great River of the West. Like any unexplored territory, the region had been home to many marvelous places never actually seen but nonetheless thought to be real. There was Cíbola, described by one purported visitor as "a land rich in gold, silver, and other wealth," with seven "great cities" whose residents were so prosperous that the women wore belts made of gold. There was Quivira, where, according to legend, common supper plates were made of gold and fish were as big as horses. And there was the land of the Madocians, a fair-skinned race thought to have descended from a Welsh prince whose reputed discovery of America in 1170 gave the British evidence that they had more claim to the continent than the Spanish. Like those places, the Great River of the West was the result more of desire than of reason. Ever since setting foot in North America, Europeans had fantasized about a waterway, or system of waterways, that would lead them neatly, safely, cheaply to the Pacific and the Orient beyond. "It must exist because it had to," wrote the historian Bernard DeVoto of this Great River. "The logic of deduction from known things required it to, and so did the syllogism of dream—both on no grounds whatever. So it did exist in personal narratives and speculative treatises, in treaties and on maps, under various names, flowing in various directions."

In 1776 a Franciscan friar named Silvestre Vélez de Escalante made a remarkable fifteen-hundred-mile trek through the Southwest, during which he crossed the site of the present-day Dinosaur National Monument between Colorado and Utah. There he came upon what is now known as the Green River. Escalante, however, gave it a different name: San Buenaventura. His cartographer, Bernardo de Miera, later placed this waterway on a map—but instead of making it head south, the direction in which the Green River actually flows, he sent it off southwesterly toward the sea. Miera never connected the Buenaventura to the Pacific, but the mapmakers who followed him were not so cautious. Before long, observed the author Carl I. Wheat, "it had become . . . a mighty stream, flowing, as befitted logic, into the San Francisco Bay or thereabout, and for many years even the most knowledgeable cartographers made this and other equally apocryphal streams flow westward from the central mountain range and from the Great Salt Lake to the Pacific."

By the spring of 1843, when John Charles Frémont launched his second big mapping expedition of the West, serious doubts had begun to emerge about whether there really was a Rio Buenaventura. The explorer Benjamin L. E. de Bonneville, for example, had returned from his extensive travels in the region claiming that no such body of water existed. Bonneville's map of 1837, as well as those of former Treasury Secretary Albert Gallatin in 1836 and Geographer to the House of Representatives David H. Burr in 1839, contained no Buenaventura, but the river remained intact on other popular maps of the period. In November 1843, having gone as far west as Fort Vancouver, near present-day Portland, the Pathfinder decided to put an end to the controversy once and for all. Instead of proceeding directly home along the Oregon Trail, he resolved to reroute his return trip to the south. He did this in large part, as he later explained, to make a final determination about "the existence of a great river flowing from the Rocky Mountains to the Bay of San Francisco." It was a typically rash and dangerous move. On no one's authority but his own, Frémont was leading his twenty-five men into a largely unknown wilderness just as winter set in. They wandered into the Sierra Nevada, expecting "with every stream . . . to see the great Buenaventura," Frémont later wrote. The higher they climbed, the more it snowed. As always with the Pathfinder, disaster awaited.

Critics often cite the journey as proof of Frémont's recklessness. His biographer Allan Nevins, for one, called it "ap-

pallingly foolhardy." Yet reading about the episode, I couldn't
help but admire what the Pathfinder was trying to accomplish.
At the time I was searching for my own waterway of sorts, a
stream of events that coursed through more than twenty-five
years of a man's life. If I were to believe Bland's lawyers, the
Vietnam War had led to his crime spree just as surely as a great
river flows to the sea. Bland's critics, by contrast, argued that his
claim of post-traumatic stress disorder was nothing more than
a convenient fiction, the courtroom equivalent of the Rio Buenaventura. Neither side, however, offered me much in the way
of evidence. And so it was that I once again found myself following the Pathfinder's lead, on a mission to separate reality
from myth.

I made a lot of discoveries on that journey. In the Bahamas I
found a map dealer named Jonathan Ramsay, who had worked
closely with Bland and seemed to know him as well as anyone
in the industry. "One thing he did talk about was the war in
Vietnam," Ramsay told me. "He had been over there and was
quite affected by it. He didn't really like having been sent over
there and he didn't really approve of the war and the way it was
conducted. And he chatted a lot about his feelings about the
Vietnam War and the people over there. . . . I can't remember
the details exactly, but I thought that he was teaching at one of
the local schools. He had the opportunity to teach the Vietnamese kids in this town, wherever it was that he was stationed,
and he said that that was one of the most rewarding parts of
being in Vietnam. I can't remember exactly what it was, a
Catholic school or a government school for orphans or what.

But he was obviously very moved by that. He felt that he was contributing something very positive."

On the East Coast, I found one of Bland's daughters by his first wife, Carol Ann Talt Bland, whom he married in 1971 and divorced in 1978. Heather Bland told me she'd never had much contact with her father. Still, she was able to recall an old family tale about his days in Vietnam.

"I heard that the war did affect him," she said. "I heard a hearsay story about one of the things that happened to him. I don't know how true this is, but supposedly while in Vietnam he saw one of his higher-up officers doing something to some female, raping her or something like that. My father supposedly beat the hell out of him and was put in solitary confinement. Rumor has it that it really messed him up."

I found no confirmation of this story during the course of my expedition. But in Hackensack, New Jersey, I found a jail docket from Bland's July 6, 1971, arrest on charges of being AWOL. It indicated that he claimed to be suffering from a "nervous disorder," for which he was receiving medical care from the Army. The docket also listed him as a "drug addict." In those same files, however, I found another docket from an arrest some six months earlier, in which Bland "denie[d] illnesses" and "denie[d] drugs."

Whatever the case, I discovered plenty of evidence about the chaotic state of Bland's life after the war. In the New Jersey State Archives I found a letter he sent to then-Governor William T. Cahill on September 29, 1972, a few months before Heather Bland was born. Writing from the Sussex County Jail in Newton, New Jersey—where he had been booked the previous day for marijuana possession—Bland begged the governor to show mercy upon his pregnant, nineteen-year-old wife. Although she was with him at the Playboy Club during his arrest, he explained, she had nothing to do with the drugs that were found in his car. Nonetheless, she too had been tossed in jail, he claimed. Insisting that he was "going crazy" because of a situation that "I would expect to happen in Nazi Germany or Communist Russia," Bland implored the governor to halt the "persecut[ion]" of his wife.

It was hard not to feel sorry for the person who wrote this letter, hard not to see him as a deeply troubled and terribly naive young man—perhaps even as a victim. And it was equally difficult to imagine how such an unsophisticated individual would someday be able to talk his way into a blue-blood subculture dominated by Ivy League types. Or at least those were my first impressions. But, as I was quickly learning, first impressions—mine and other people's—about the map thief were often misleading. In police and court files in California and Florida, I found evidence of a Bland very different from the author of that forlorn missive. This was Gilbert Bland, inventor of imaginary creatures, exploiter of the system, professional con man—a persona that may have already been present in that jail cell in New Jersey. Files from the Sussex County case show that Bland, aka James J. Edwards, never showed up in court to face his drug charge, and that a warrant for his arrest (apparently still open to this day) was issued on June 22, 1973. (Those same court files, as well as local police records, indicate that, despite

Bland's implication to the contrary, no charges were ever filed against his wife.) The next time this alter ego surfaced was in California, where, in September 1973, Bland, aka Jason Michael Pike, aka Jack Arnett, was arrested on grand theft charges stemming from a credit card fraud scheme. And, unlike the Bland of a year earlier—a man so seemingly ignorant of the judicial process that Governor Cahill had to advise him to "give consideration to retaining legal counsel"—this one was a shrewd manipulator of the courts. Facing a maximum penalty of ten years in prison and a five-thousand-dollar fine, he agreed to a plea bargain and was given five years' probation. That, however, did not seem to slow his career as a scam artist. On December 30, 1975, federal agents nabbed him in Florida on charges stemming from his use of false identities to rip off unemployment benefits. This time, his long string of luck in the courtroom had finally run out. Even with another plea-bargain deal, he got a three-year sentence.

In files from that case, I found a letter he wrote shortly after arriving at the federal prison in El Reno, Oklahoma. In that note Bland begged the judge who had sentenced him for help, explaining that other prisoners had unjustly accused him of " 'snitching' as they call it, a crime in here punishable by death." He elaborated:

> *Everyone in here is associated in "clicks" or gangs. Since I am an outsider to these "clicks" I live a fearful existence. I pray before I go to sleep at night that I'll wake up the next morning. It's common knowledge among the inmates here that for [$]30*

worth of comissary [sic] items, a person could be arranged to
be killed. Weapons are not hard to come by in here. I know I am
obtaining a paranoia complex because I fear it's only a matter
of time before something happens to me. . . . I don't know how
long I can bear this pressure. I don't know who to talk to. My
counsellor said if I'm going to be in a fight to choose my own
place and time and make a lot of noise so the guards can come
and break it up before I'm hurt bad. I don't know if I can fight.
I'm not a violent person and I thought I was being sent to a
place with non-violent people, but that's not the case here. . . . I
don't think this pressure is the punishment you had in mind for
me. It's very hard to live in fear all the time . . .

Were these the words of a shell-shocked veteran who once
again found himself trapped in a windowless box, dreading
sounds in the darkness? Or were they simply the latest fictions
of a calculating flimflam man? I could never be sure. Nor could
I judge whether Gilbert Bland believed his own words when he
wrote that same judge eleven months later to vow, "I have no
intention of ever committing an illegal act again."

THE PATHFINDER SURVIVED THE SIERRAS, BUT JUST
barely. When he and his men became bogged down in snow as
deep as they were tall, their Indian guides deserted, and one
starving member of the expedition, suffering from hallucina-
tions, wandered off into the wilderness, never to be seen again.
The cartographer, Charles Preuss, lost from the main party for

three days, was forced to eat ants to survive. Yet when the dazed adventurers finally stumbled down into the Sacramento Valley, at least one thing was clear in their minds. As John Noble Wilford wrote in his book *The Mapmakers:* "Frémont's march south to California eliminated the Buenaventura from the map."

My own expedition was not so successful. I found a few promising streams and many dry riverbeds but discovered no unmistakable channel linking Bland's experience in Vietnam with his cartographic crime spree. Yet neither could I conclusively discount it as a myth. By now I had learned that sometimes you don't conquer the wilderness even when you have a guide. I would add many new coordinates to my map in the months to come, but this particular one would always resist my geography. In the end I was left to ponder one last story about the Rio Buenaventura. For centuries mapmakers theorized that the Great River of the West and all the other major rivers on the continent had their headwaters in the same place, a so-called height of land. At the beginning of the nineteenth century, this concept assumed a brief life as cartographic fact when Zebulon Montgomery Pike claimed "no hesitation in asserting" that, on a journey to present-day Colorado, he had come within a short distance of this "grand reservoir of snows and fountains." He should have known better. The true source of a thing is never so easy to find.

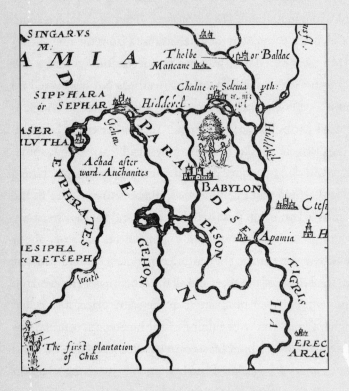

THIS MAP, WHICH ACCOMPANIED A
1614 EDITION OF WALTER RALEIGH'S
HISTORY OF THE WORLD, LOCATES
THE EARTHLY PARADISE IN
MESOPOTAMIA.

CHAPTER 9

The Waters of Paradise

*F*LORIDA HAS ALWAYS BEEN A LAND OF new beginnings. Even the guy who put the place on the map came in search of metamorphosis. As a Spanish functionary in the West Indies, Juan Ponce de León had heard local legends about an island with "a spring of running water of such marvellous virtue, that the water therof being drunk, perhaps with some diet, makes old men young again," in the words of one sixteenth-century chronicler. Funny thing: European myths foretold a similar fountain, usually set in some far-off enchanted garden—sometimes Eden itself. You may remember, for example, that Sir John Mandeville, the enigmatic fourteenth-century figure we examined earlier, claimed to have visited a "noble and beautiful well, whose water has a sweet taste and smell, as if of different kinds of spices." Never one for understatement, Mandeville claimed that "he who drinks of it seems always young. They say this water comes from the Earthly Paradise, it is so full of goodness." Could it be mere coincidence that two legends half a world apart were so similar? Apparently Ponce de León didn't think so. Feeling past his prime at age fifty-three, he set sail from Puerto Rico in 1513, reportedly to search for gold and a cure for *el enflaquecimiento del sexo,* the debility of sex. He never located the Fountain of Youth, but he did stumble upon an even greater marvel: North America.

And, in a curious way, Ponce de León's quest has never ended. To this day Florida remains a mecca for those trying to make a new start. Possessing one of the country's lowest proportions of locally born citizens and highest rates of population growth, the state teems with wealthy Yankee retirees, ambitious Caribbean immigrants, and a hodgepodge of others who come in search of a new lease on life. The sense of rootlessness is as thick and ripe as the midsummer air.

Gilbert Bland, too, began his life all over again in Florida—not once but twice. The first time came upon his release from the federal penitentiary in late December 1977. After being discharged from a Florida halfway house in March 1978, Bland settled down in the Fort Lauderdale area, where, in May of that year, he went to court to divorce his first wife, Carol Ann Talt Bland. It would turn out that he was breaking not just with one person but with his whole past. According to Heather Bland, a daughter by that first marriage, her father rarely contacted her or her sister, Melissa, as they were growing up. "He's got two children that he totally abandoned and couldn't care less about," she told me.

Nonetheless, Bland appeared to turn his life around, at least for a few years. He married again, and he and his new wife, Karen Bland, had two more children. After receiving an associate's degree from Broward Community College in Fort Lauderdale and studying for three quarters at Florida Atlantic University, he moved to Maryland, where he worked for Allied Signal Corporation for five years during the 1980s. He may have also developed an interest in antiques during this period. Statements Bland later made to other dealers, as well as to police, indicate that his fascination with maps grew out of a more general interest in antiques, apparently including old stocks and bonds, as well as rare books. This notion was later confirmed by one of his attorneys, who wrote: "His interest in maps developed through his prior interest in and collection of older books, some of which contained maps in them."

In 1992 he and his wife opened their own computer training and leasing firm, Pacific Data Systems, in the town of Columbia. The firm, according to Maryland state records, was owned by Karen Bland—because, I would imagine, her husband's fraud convictions did not make for a particularly good credit rating. The couple apparently worked out of their home, since the company's mailing address was a postal box at Parcel Plus, a mail services outlet in Columbia. I could not find out much about the specific nature of Pacific Data's operations, but court documents from a later case offered a few fascinating details. One of Bland's "major customers," according to the documents, was an organization he would soon be dealing with in a vastly different capacity: the Federal Bureau of Investigation. The documents, drafted by Bland's attorneys, asserted that he "made a number of sales" to the FBI office in Maryland, where he "trained agents in the use of such equipment, and . . . assisted in a technical capacity with computers used during and for surveillance operations."

The FBI, prompted by a request I filed under the Freedom of Information Act, confirmed that a Pacific Data Systems of Columbia, Maryland, did indeed perform unspecified work for the Bureau in 1993. But even with a government contract, the firm was apparently far from a rousing success. Paul R. Thomson, Jr., one of the map thief's attorneys, would later assert in court that a changing market in the early 1990s made the concept of computer leasing "no longer . . . economically feasible." As Bland's business dried up, he and his wife decided to move to Florida to be with her mother, who "had been recently widowed," according to Thomson.

Bland departed from Maryland sometime in late 1993 or early 1994, leaving behind his usual share of unfinished business. At the Parcel Plus in Columbia, for example, an employee told me that Bland had pulled up stakes without paying his bill—and without leaving a forwarding address. It was yet another clean break from the past.

In February 1994, the Blands opened Antique Maps & Collectibles, Ltd., in the Gardens, a sleepy little office and retail complex located in the Fort Lauderdale exurb of Tamarac, whose open-air courtyard echoes not with the bustle of shoppers but with the listless gurgle of fountains that line the empty walkways. Perhaps, as he stared out at those fountains, Gilbert Bland thought that he had succeeded where Juan Ponce de León had failed, that he had discovered a way to wash away the past and reinvigorate a sense of potency. Perhaps, beginning a new career in a new town just months shy of his forty-fifth birthday, he felt almost young again.

Or perhaps he just wanted to be left alone. "Basically, the man stayed to himself," said Jan English, a bartender at the Beverly Hills Cafe, an eatery at the Gardens. Added one employee of a business that faced Bland's shop: "The place was basically always empty. We were sitting here one day thinking, I wonder how he makes money? And then we were wondering, Who would be interested in those old maps?"

The choice of location for the store must have seemed peculiar indeed to casual observers. Lost amid South Florida's vast formless expanse of strip malls and subdivisions—the kind of landscape where "everyplace looks like noplace in particular," in the words of James Howard Kunstler, author of the anti-sprawl tome *The Geography of Nowhere*—the Gardens was about the last spot you might expect to find an antique maps shop. Then again, this was one antique maps shop that had no desire to be found.

THE MAP DEALERS DID NOT TRUST ME; THAT MUCH WAS obvious from the start. When I called them they would often beg out of the conversation, some politely explaining that they were simply too busy to talk in the foreseeable future, others bluntly telling me to mind my own business. When I visited them at the 1998 Miami International Map Fair, once again following in Bland's footsteps, their eyes would sometimes dart to the floor as I approached, or their conversations would come to a sudden halt, rising again in whispers after I walked past. In some cases it took me years to gain their confidence, and even then they would usually speak to me only on the condition of anonymity.

At first I attributed their guardedness to simple embarrassment. Many of them had done a considerable amount of business with the now-notorious map thief, a fact that they hoped their colleagues would forget and their customers never learn. But as time wore on I began to see a more subtle factor at work. The world of antique maps is a very small and genteel one. Although their shops are located all over the globe, the dealers regularly attend the same gatherings (including the Miami map fair), read the same journals (such as the glossy *Mercator's World* or the scholarly *Imago mundi*), and keep in touch on the same Internet forums (most notably the Map History Discussion List), where they join librarians, scholars, and collectors in exchanging esoterica, debating map controversies, and trading cartographic jokes. *(What did the mapmaker send his sweetheart on Valentine's Day? A dozen compass roses.)* I was a complete stranger to this punctilious subculture. Even so, I had the feeling that in normal times I might have been made to feel welcome. These, however, were not normal times. Three years earlier, another outsider, someone the dealers had known no better than they knew me, had come into their midst asking for their trust. They had made the mistake of giving it to him. "It's a very close community," explained F. J. Manasek, a well-known Vermont dealer and industry observer who wrote *Collecting Old Maps*. "We're all friends, even though we compete in business. There's a lot of honor, which is probably why Bland could gain such easy entrée."

"The map trade is basically a handshake business," added George Ritzlin, another longtime dealer, based in Illinois. "People trust each other. Substantial amounts of valuable material travel thousands of miles based on a telephone call, without any other supporting documentation. And this is why Bland was able to be so successful. He was incredibly brazen by putting himself up to be a dealer. People just assumed that anyone acting as a dealer had obtained material honestly. Thieves don't normally work that way. Normally they are very surreptitious."

Having been burned once, the dealers now had little intention of exposing themselves to another outsider, especially one determined to open old wounds. I had not counted on them viewing me this way. Somehow I had convinced myself they would look upon me as an ally. It took a very long time to realize that when they glanced at me with those disdainful eyes, the man they saw looking back was Gilbert Bland.

ONE OF THE FEW DEALERS WHO WOULD DISCUSS THE matter openly was Barry Lawrence Ruderman, himself a relative newcomer to the trade. A bankruptcy lawyer by profession, the thirty-eight-year-old Ruderman discovered antique maps in his late twenties, when, during a ski trip to Taos, New Mexico, he "got bored" one day and wandered into a map shop. It was, said Ruderman, "love at first sight," and his passion soon developed into what he and other map junkies sometimes call cartomania.

"Cartomania is a sickness," he once told me. "It's obsessive. Once you're in up to your ankles, you want to be in up to your knees; once you're in up to your knees, you want to be in up to your waist. I like to think that it's sort of a beautiful sickness, because all human beings need things that stimulate them intellectually and drive them to passion. But the secondary aspect is that many of us spend insane amounts of time dedicating ourselves to map collecting. It's a twisted pursuit. But where's the problem in that?"

In 1992 Ruderman turned his "sickness" into a sideline vocation, opening Old Historic Maps & Prints, a La Jolla, California, venture that sold rare maps via mail and the Internet. Two years later he received a catalog from another fledgling map retailer, this one based in Tamarac, Florida. "It was semiprofessional looking, nothing real fancy," Ruderman remembered. Although the catalog contained "a bizarre mix" of materials, including a lot of worthless junk, Ruderman was intrigued by some of the offerings, especially maps of the American West. He would become one of Gilbert Bland's earliest clients.

But make no mistake: the fact that Ruderman was one of the only dealers candid enough to publicly discuss his dealings with Bland in no way means that he was the only one to have *had* such dealings. Indeed, the majority of Gilbert Bland's customers appear to have been dealers, including some of the most respected names in the trade. The map thief and his wife did little to encourage a walk-in business at their Florida store, which, like the brand-new home they bought for $151,400 in nearby Coral Springs, was owned by Karen Bland. Instead, they concentrated on developing a long-distance clientele, advertising in international trade publications and sending out catalogs. In retrospect, they seem to have penetrated the world of antique maps with startling speed and effectiveness.

Bland's inwardness worked to his advantage. Someone else in his position might have tried to ingratiate himself, make friends, become an indispensable member of the group—in short, to do what con men do: gain people's confidence. But whether by intent or natural inclination, Bland was less of a con man than an *un man,* inducing unmindfulness, lulling people into believing he was simply not worth much thought one way or another. As one dealer later put it: "Mr. Bland was bland. He looked bland, he sounded bland, he acted bland. There was no personality: nothing there."

Along the same lines, Ruderman remembered him as being "pleasant, not a fabulous conversationalist, but nice enough, friendly enough, a tad bit socially awkward, maybe. Just your average Joe. . . . There was nothing to raise an eyebrow about, other than the fact that here was a guy with a large cache of material whom nobody had dealt with before."

And, luckily for Bland, many in the industry were then in

need of a just such a "large cache." The map business was undergoing something of a sea change when Bland arrived on the scene. With big numbers of new collectors and traders entering the market, the demand for antique maps was higher than ever. For obvious reasons, however, the supply had remained pretty much stable. That meant not only higher prices but fewer maps in active circulation. This was no problem for major players like Graham Arader, who could afford a huge overhead. But for midsize and smaller dealers, the capital that had once bought five or ten maps now might purchase only one—*if* that item was even available. Hard-pressed to maintain an adequate inventory, these dealers found themselves scrambling for new sources of material. It was no wonder that word spread quickly about a Florida store with an incredible supply of low- to mid-priced maps.

For his own part, Ruderman "bought actively with Bland for probably six or eight months." On occasion he entertained doubts about his business associate, in part because he seemed to know so little about maps. One time Ruderman even "sort of cross-examined Bland" about the provenance of his materials. Bland replied that he and his wife had been involved in scripophily—the collecting of old stock certificates and bonds—and had incidentally been accumulating old maps. "That was an acceptable answer," said Ruderman, "because frankly there are two or three respected dealers who fit that general MO." In the end, Ruderman concluded, "Gil passed the smell test."

FOR OTHERS, HOWEVER, THE ODOR EMANATING FROM South Florida had become increasingly foul. Among those who bought items from Bland were a pair of dealers who have operated a respected map store for several decades. The partners, who spoke on the condition of anonymity, became increasingly suspicious of Bland, eventually severing ties with him and privately informing other dealers of their concerns. We'll refer to them as Once Bitten and Twice Shy.

Their first contact with Gilbert and Karen Bland came in May 1994, just three months after Antique Maps & Collectibles had opened. "We heard about him from another dealer whom we're quite friendly with," recalled Once Bitten. "During a phone conversation, the fellow said, 'Well, this couple in Florida seems to have a lot of material. You probably want to get in touch with them.' So that's what we did. At first the Blands were very pleasant. . . . They didn't seem terribly knowledgeable, but they seemed to want to learn. So we tried to be helpful, just as people were helpful to us when we started."

Thus began a short-lived relationship in which the veteran dealers functioned both as customers and as mentors to the Blands, educating the newcomers on what they were doing wrong and how they might correct it. There was much to discuss. Among other problems, the Florida firm's early catalogs contained a number of descriptions that were "ignorantly inaccurate," according to Twice Shy. For example, the Blands misidentified the mapmaker of an eighteenth-century French map of China as Échelle. They had apparently found this word in the cartouche, the part of the map where the cartographer's name usually appears. Ah, but Monsieur Échelle, he was no mapmaker! Poor soul, he did not even exist! *Échelle,* the veteran partners dutifully informed the Blands, is the French word for "scale."

For their part, the newcomers always seemed grateful for such enlightenment, however humiliating. In one obsequious letter Gilbert Bland praised the partners' "eminence and experience in the field of cartography/map collecting," vowing to "accept and incorporate" their ideas "into my own limited knowledge of this business."

Karen Bland also had regular correspondence with the pair. It should be stressed that no criminal charges were ever filed against the map thief's wife; even so, observed Once Bitten and Twice Shy, she seemed to be intimately involved in the day-to-day operations of the store. "She knew as much as he did, that was my impression," said Once Bitten. "My feeling was that if we wanted to engage in any sort of business with them, we could talk to her as readily as him. I thought that she was an active part of the business."

Despite the cordial relationship, it did not take long for Once Bitten and Twice Shy to notice that some of their most important advice was being ignored. They became particularly frustrated with the Florida couple's continued failure to adequately describe the materials for sale in their catalogs. This was not a mere matter of snotty etiquette: the Blands were listing items in a way that implied that they were in top condition when in fact many of them were in "wretched shape," according to Twice Shy. "We'd send stuff back and say, 'Listen, we're not being picky. Here's what people want and here's how you should describe it.' Then the same items would appear on the next list with no correction at all. And Gil would always blame it on the computer—somehow it didn't catch in the computer when he made the correction."

Trying hard to give the newcomers the benefit of the doubt, Once Bitten and Twice Shy offered advice about which software might solve the problem—unaware that, in another, not-so-distant life, the Blands had been computer consultants. After a number of frustrating incidents, however, the veteran pair's patience finally wore out. "I remember one of the things they sent to us actually was missing a chunk," said Once Bitten. "We sent it back. And in the next catalog, there it was again without any description indicating it was incomplete. . . . At that point, it was pretty clear that they were being deceptive. When someone is told in no uncertain terms that a map is incomplete, is missing a piece, and should not be described in the way they were describing it—and then chooses to do it again, it's hard to think of another excuse for their actions. They weren't stupid."

By the time they broke off ties with Antique Maps & Collectibles in late 1994, Once Bitten and Twice Shy had begun to have other suspicions about Gilbert Bland. They were troubled, for instance, by his "bizarre" style of negotiation. Bland's prices were fairly cheap to begin with, yet he always seemed willing to take a lower offer. "He never counteroffered," said Once Bitten.

"He'd just say, 'Well okay.' . . . In retrospect, we know that he didn't care so much about the price as long as he got *something*. It was all profit."

Worse, the partners had growing doubts about where Bland was getting his materials. At first, explained Once Bitten, they had assumed that he "lucked into something. Maybe he had bought an antique shop and there was a pile of stuff in the basement or something of that sort." But over time they began to suspect more ominous origins.

On the day I interviewed the partners at their store, they took out some of Bland's old catalogs, shaking their heads as they recalled the way his inventory mysteriously expanded to include multiple copies of fairly rare items. "It was frustrating to comprehend how this guy, whom I considered a classic jerk, was getting access to materials that we who had been in the trade for so long had to wait for years to pop up," said Twice Shy.

Because they had no hard evidence against Bland, the partners did not go to the police or denounce him in a public way. They did, however, begin to share their hunches with other dealers. "It was not like we called everybody up and said, 'Wow! We think this guy is rotten!' " explained Twice Shy. "People in this trade chat each other up all the time. We're as gossipy as a bunch of old ladies. And Bland's name would just come up in conversation."

They soon discovered that others had similar concerns. "As those of us old-timers in the trade started questioning how this guy was getting duplicates of some more interesting material, I was speaking with another dealer and I said, 'If I had money, I would hire a private detective to follow Bland,' " said Twice Shy. "I was *that* suspicious."

But if some dealers were beginning to steer clear of Gilbert Bland, others were becoming increasingly enamored of his growing stock of materials, which by now included a number of older, rarer, and more expensive items. His reputation on the rise, Bland and his wife made high-profile appearances at the two big industry conventions of 1995—the Miami International Map Fair in February and the International Map Collectors' Society fair, held in San Francisco during October. "Bland had a major presence at both fairs," recalled James Hess, who owns the Heritage Map Museum and Auction House in Lititz, Pennsylvania. "He was putting himself out there with the major dealers."

Barry Lawrence Ruderman, who had dinner with Bland at the San Francisco event, added: "Most of all, he was interested in being a wheeler-dealer. He was looking for big buys. He was definitely crunching numbers a lot more than he was learning maps."

"It got to the point," recalled one respected antiquarian, "that dealers would be saying, 'My goodness, maps of City X have been selling rather well. Do you have any maps of City X?' And Bland would say, 'Let me check and I'll get back to you.' And the very next week he'd call and say, 'Why yes, I just happen to have a map of City X.' "

Which, of course, is exactly why the dealers should have been wary. The fact that many continued to do business with him right up until his misadventure at the Peabody Library says less about the nature of the map trade than it does about that oldest of human afflictions, temptation. Almost no one I interviewed believed Bland's customers had firsthand knowledge of his crimes. More likely, the dealers were driven by what antiquarians sometimes describe as "a need not to know." In the words of Once Bitten: "I'm sure a number of people closed their eyes. It's very easy to do when there's a chance to make money."

*F*OR GOD DOTH KNOW THAT IN THE DAY YE EAT THEREOF, *then your eyes shall be opened, and ye shall be as gods, knowing good and evil.* Such were the words laid down in ancient times about what transpired "eastward in Eden." Yes, wondered the people of later epochs, but where, exactly, was this Eden? Paradise was lost, quite literally—and for centuries theologians, scholars, pilgrims, and mapmakers were obsessed with finding it.

Earlier civilizations, from the Sumerians to the Greeks, had established a lush and fragrant garden as the basic setting of Paradise—a word that comes from the Old Persian *apiri-daeza,* an orchard surrounded by a wall. The Hebrew Scriptures fleshed out those myths, offering a seemingly solid clue as to this enchanted garden's whereabouts. According to the second chapter of Genesis, a stream flowed through the Earthly Paradise, then branched into four rivers—the Pishon, later thought to be the Indus, the Ganges, or the Danube; the Gihon, or Nile; the Hiddekel, or Tigris; and, finally, the Euphrates. This description, however, created more questions than answers—not the least of which being how four rivers whose headwaters were apparently so very far apart could possibly spring from the same source. Some early Christian scholars argued that such concerns were beside the point, since the Bible's description was merely figurative. But many medieval theologians insisted that the Garden was an actual place on Earth, one whose rivers flowed underground before reaching their apparent starting points. The learned St. Isidore of Seville, for instance, located Eden in the Far East, surrounded by a "wall of fire whose flames rise as high as heaven," making it "barred to humanity." Reflecting such ideas, the *mappae mundi* of the Middle Ages usually show the Garden at the easternmost edges (or top) of the world, sometimes as a walled-off region on the mainland, sometimes as an island above "farthest India." Such concepts remained pretty much standard until the Age of Exploration, when new possibilities arose, especially in the New World, where Christopher Columbus claimed to have seen "great evidence of the earthly Paradise" near the Orinoco River. The hunt would go on for centuries, although its focus would eventually shift from where Paradise *still was* to where it *had been*. Sir Walter Raleigh—who argued that the Garden itself could no longer be found, inasmuch as "the flood, and other accidents of time" had reduced Eden to the state of ordinary fields and pastures—placed its original location in Mesopotamia. Others put it in Armenia or Palestine or Africa or Ceylon. As late as the nineteenth century, a British general reported absolute proof of Eden's existence on Praslin Island in the Indian Ocean. Charles

Gordon, a war hero and religious zealot, was convinced that a species of palm on the island was none other than the Tree of Knowledge of Good and Evil. After all, he noted, the plant bore fruit of both sexes, the male resembling a phallus, the female "shaped . . . when opened like a [woman's] belly with thighs." How could such suggestive vegetation be anything but proof that Adam and Eve had lived on the island?

These days, of course, not even the truest of true believers would dare to put Paradise on a map. Yet despite the cynicism of our age, we humans have not lost our urge to quest after that place of perfect contentment, never quite finding it but never quite giving up hope, sometimes drawing so near that we can almost smell the faint sweet scent of its blossoms or spy the distant glimmer of its waters. I still don't know precisely what motivated Gilbert Bland to open that store in South Florida. His detractors argue that his intentions were evil from the start, pointing out that James Perry's visits to university libraries began just a few short months after the store opened. True enough, yet somehow I'd like to think Bland's motives were more complex. My hunch—or maybe it's just my hope—is that, in the beginning at least, his desire to start his life anew was sincere. But however the saga at the strip mall began, it ended as tragicomic trash-culture allegory: *And the sinner and his wife were driven forth from the Gardens retail center, and their sorrows were greatly multiplied, and they knew then that they were naked before the Almighty Media Juggernaut.*

The Fall of Bland came in mid-December 1995, a week after he was detained at Johns Hopkins. With the University of Virginia's Thomas Durrer and the FBI closing in on him, Bland had returned to Florida. At some point between the afternoon of December 14 and early morning of December 15, he cleaned out his store and left the Gardens forever, reportedly leaving a note for his landlord that said, "See you later."

"I came in and a lot of the maps were gone," said Laurie Bregman, a tenant at the Gardens whose business was just across from Antique Maps & Collectibles. "I thought that maybe he was just remodeling or that maybe he'd sold a lot of stuff. But it turned out that he had emptied the place in a middle-of-the-night kind of deal."

Although Bland had managed thus far to escape arrest, he had not escaped attention. A number of major media outlets, including the Baltimore *Sun* and Associated Press, had already run stories about the crime spree; many more, including *The Washington Post*, the *Chicago Tribune*, and National Public Radio, would soon follow.

The news spread quickly through the antique maps industry, prompting one of two sharply divergent reactions among dealers. For some, it was a fait accompli. "I was completely unsurprised when the story broke," remembered Twice Shy. For others, however, it was a complete shock. "My jaw dropped," said Jonathan Ramsay, the owner of a map and print business in the Bahamas. "I mean, the guy was straight as an arrow. When something like this happens, you say to yourself, 'Wow, I just don't understand human nature.' "

Ramsay, who had both a business and a social relationship with Bland, heard about the map thief's legal troubles when a friend from Florida faxed him a newspaper story. "I called my friend up and I said, 'You've got to be kidding me!' Just then I heard a noise in the shop and I turned around and there were these three guys standing there. And I said, 'Yes? Can I help you?' And one of them said, 'I'm an FBI agent.' He'd been standing right behind me as I talked on the phone. He said, 'Obviously I can hear from your phone conversation that you've heard the news.'"

Another dealer who bought maps from Bland and met him face-to-face shared Ramsay's surprise. "Bland was the most soft-spoken and considerate guy," he said. "It was like a contradiction. On the phone and in person, he was so quiet, and then on the other hand, the crimes he committed were incredibly nervy. I guess he was a hell of a con man."

Those who lived in the same upscale subdivision as the Bland family could be forgiven if they, too, were taken aback by the news. The four-bedroom, two-and-a-half-bath, Italian-style house, part of a development called the Classics at Kensington, projected an air of respectability and affluence. But Gilbert Bland's life had been full of misleading facades, and this one was no exception: The family's financial picture was nowhere near as rosy as it appeared from the outside. On November 3, 1995—a little more than a month before Bland's brief capture in Baltimore—Karen Bland, the owner of record for both the house and the store, had declared Chapter Seven bankruptcy. Court documents show that she owed more than forty thousand dollars in credit card debt alone.

Bland's lawyers would later argue that it was the failure of his computer leasing firm in Maryland that led to these financial troubles and eventually to his crime spree. But that might not have been the only factor. "I'll put another scenario in front of you," said Jonathan Ramsay. "He came over here [to the Bahamas] about four or five times—and he liked to gamble. He'd say to me, 'I'm over here on a gambling junket.' He would come in to see me and then he would go off to the casino."

Given Bland's apparent propensity for risk, an infatuation with gambling would hardly seem out of character. But whether to pay off casino debts or for some other reason entirely, he clearly found himself in desperate need of fast cash—and knew a way to get it. "He found an easy avenue to make some quick money, and he really overdid it," said Lieutenant Detective Clay Williams of the University of North Carolina Department of Public Safety, one of several law enforcement agencies that was by now getting involved in the case. "He got in way over his head. It became addictive. I don't think he had any conception of the federal charges that could come down on him."

Come down they did—but slowly, very slowly. While news of the crime was breaking all over the country, the FBI's attempts to get a search warrant for Bland's store had gotten bogged down in red tape. "A search warrant has to be originated in the same jurisdiction that it's going to be executed," explained Special Agent Henry F. Hanburger of the FBI office in Columbia, Maryland. "And timing was terrible with the

Christmas holidays and the absence of people at work both on the prosecutor's side and our side. . . . We just couldn't get the needed prosecutor to say, 'Damn right, let's get a search warrant.' That cost us a little bit."

In fact, the delay would end up costing investigators a great deal—and prosecutors even more. It was not until a day or two *after* Bland had cleaned out his store that FBI agents tracked him down in Florida. And it was not until two weeks later, January 2, 1996, that the map thief turned himself in to local police. At long last Bland was in custody. Now it was his maps that were nowhere to be found.

CHAPTER 10

The Joy of Discovery

*T*HE ADMIRAL SITS DOWN TO WRITE A LETTER. *From outside his cabin come sounds so familiar he does not even hear them: the creak of wood, the slap and flutter of sails, the hoarse lurch of the sea. He is on a ship called the Niña, heading east. As he leans over the page, his blue eyes narrow with excitement, and he indulges himself in a satisfied smile. The letter is dated "February 15 of the year 1493," and is addressed to a Spanish official. It contains thousands of words, only two of which really matter:*

I discovered . . .

The admiral does not know exactly what he discovered; he will go to his grave not knowing. (For now, he vaguely describes it as a great many islands inhabited by people without number.) Nor does he understand how completely his discovery will change the world, or how fast. Within months copies of this same letter will be circulating

Island of Lost Maps

An allegorical commemoration of
Columbus's discoveries in the
New World, from a 1621 book.

throughout Southern Europe, published in a dozen editions and in three languages. Within months the admiral will be a celebrity, cheered by throngs of well-wishers and feted by kings, queens, courtiers, and clergymen. Within months other explorers will be laying down plans to retrace the admiral's route. They, too, will make discoveries, but none of those will compare with his. The admiral has crossed an uncharted sea, visited an unmapped world: even though people may have spoken and written about these lands, all was conjecture, nobody actually having seen them. *To seek and to find—this has been his dream almost since childhood. ("He was charmed by the afterglow of legends that flashed from the unknown," the biographer Gianni Granzotto would later observe. "He began to fantasize about them, and they became the constant subject of his thoughts and dreams; he thus learned to travel the invisible roads that lay between wisdom and exaltation.") To seek and to find—and now, he writes, it has come to pass:* the conquest of what appears impossible. *The admiral has overcome all obstacles—the danger, the hardships, the terrifying uncertainty, the fact that learned and powerful men said it could not be done. To seek and to find—it is his triumph alone, he knows that, but he also knows that it means nothing unless it is shared, transformed from event to story and then to legend. That is why he is returning with specimens: gold, amber, cotton, herbs and chilies, caged parrots and other exotic birds, even human beings:* I bring Indians. *That is why he is writing this letter. That is why the still-wet ink glistens with these words:* The whole of Christendom should rejoice and make merry . . .

I TRY TO MAKE IT A RULE, DURING SOCIAL CALLS, TO RE-
frain from unsolicited comments about the attire of my hosts.
Then again, not every host comes to the door dressed in South
Seas islands.

"Your shirt," I blurted out, "is a map."

This unusual form of greeting did not seem to faze the
white-haired man who stood on the other side of the welcome
mat, studying me with a Cheshire-cat grin.

"Oh sure," he replied cheerily. "I've got map shirts, map ties,
map just about everything. Come on in."

And so began my visit to the extraordinary home of an even
more extraordinary person. For the next few hours my host
would lead me on a wide-ranging and high-spirited review of
the history of mapmaking, an astounding chunk of which happened to be hanging on the walls of his large house. Maps in
the bedrooms. Maps in the lounge. Maps in the dining room.
Maps in the living room. Maps in the hallways. ("There are a
few rooms that *don't* have maps," he observed, his boyish eyes
twinkling from behind a pair of big square glasses. "Fewer all
the time.") Maps on shelves and in cabinets and in cases. Maps
spread out on tables. But not just any maps—these were some
of the rarest and most significant maps ever printed, a collection of such scope and import that no less an expert than
Graham Arader had described it as "unbelievable." And Arader's
opinion of my host? "Oh, he's a giant! A giant collector!"

My visit here was yet another detour off Interstate Bland. Collectors in general had played only a small part in the map thief's saga; my host, to the best of my knowledge, had played none at all. Yet I had come to believe my research would be incomplete until I understood the nature of collecting. I knew, of course, that the collecting impulse was, in a general way, what created the market for Bland's stolen maps. I also thought this urge might help to explain why the map thief had built up his illicit hoard in the first place—and why he had then stashed it away from police. But I had a vague notion that it had an even deeper, if less immediate, significance to the story. I did not know then that my search for answers would turn into a journey of self-revelation. I simply had a hunch that an investigation was in order—and what better place to start than this bloated temple of cartographic wonders?

The house, I should note, was extremely well-secured—yet because its contents are literally irreplaceable I have decided to publish neither its location nor the name of its owner. Let's just call our "giant collector" Mr. Atlas, not only because he seemed to hold the whole universe aloft on his walls but because the house itself struck me as a kind of atlas—multiple views of the world making up a single worldview, an open book of one man's obsession.

And make no mistake: Mr. Atlas *is* obsessed. If you couldn't figure that out from his shirt (which, he complained, was not an accurate reproduction of any specific map but a "cobbled together" representation of "Cook's voyages of the South Pacific—though I've got lots of others"), you might have guessed it from the portraits hanging in the front hall. No, those bearded gents were not his ancestors. The pair in that double portrait, each of them holding a globe, were the legendary Dutch cartographer Gerard Mercator and Jodocus Hondius, the Amsterdam engraver and map seller whose atlases made Mercator's work famous. And hanging beside them, that fellow with the tiny little head perched atop the great big pleated ruff was none other than the great Abraham Ortelius, a name and face so familiar in the Atlas family that, recounted our giant collector, when "we visited a map dealer about a year ago, and my grandson saw a copy of the Ortelius portrait there, he just couldn't understand how it could be Ortelius because Grandpa already had Ortelius in his house."

In short, it was safe to conclude that Mr. Atlas suffered from a severe case of the malady described on a poster in his basement stairway:

WARNING

OLD MAP POX

Highly contagious!

There is no known cure!

Infection is characterized by dizziness and sweaty palms

when reading old map catalogues.

This is often accompanied by apnea and lust.

Additional pathognomonic signs are:

loss of free wall space and a self-induced poverty.

If infected, do not see your doctor, but
seek aid from an antiquarian map dealer who,
while unable to effect a cure,
can provide symptomatic relief.

But I would be loath to create the impression of Mr. Atlas as some sort of zoned-out eccentric. Articulate, dignified, and keenly intelligent, he was a man whose passion for maps was surpassed only by his knowledge of them. Retired from the business world, he now worked at an academic institution, where he researched and occasionally wrote about matters of cartographic history. He and his wife also helped to support various library acquisitions, as well as sponsoring a lecture series and occasional museum exhibitions. And, of course, his home was nothing less than a museum in its own right, painstakingly assembled over the course of some thirty-five years. As he led me on a tour of what he often referred to as "the treasures," I felt privileged not only to be in the presence of such marvelous artifacts—a few of which could literally be found only within those walls—but to get to know the man who had assembled them.

With a calm half smile fixed steadily on his full-lipped mouth, Mr. Atlas had the air of a slightly distracted sage—one happy to impart his wisdom upon a willing, if sometimes embarrassingly naive, pupil. "How many people can I share this with?" he said, gesturing to the walls. "Most people don't care." And so, in the slow cadence and indulging tones of a particularly patient grade school teacher, he began to speak about his maps and his life.

"Ever since I can remember I've collected something, whether it was seashells, minerals, postcards, stamps, coins—all at a child's level mostly," he explained. "I don't know whether you're just born with an interest in collecting or what. I guess either you're a person who accumulates things or one who gets rid of things. For example, if I have a file drawer, I'm not going to weed it out until it won't close anymore."

He discovered old maps while working as an investment counselor in the early 1960s. A business associate's office had recently been redone by an interior decorator who thought maps "created the right image." Two of the artifacts had thus appeared on the man's wall. "I think one was France and the other was Central Europe," Mr. Atlas said. "I don't even remember them exactly. But I said, 'Wow, those really are neat, Bob. Where'd you get those?' Well, he just mumbled something; he couldn't have cared less about maps. But for me, something clicked. I'd always been interested in art history and geography, and I thought, Gee, here's a product that brings both those interests together. And clearly it must be affordable also, or it wouldn't be on this guy's wall. So that got me started. I called the decorator and found out where the maps came from, and that led me to one of the most important dealers in the country. It was like a whole new thing opened up.

"When I started with maps," he added, "it was a miscellaneous assortment that had only a personal connection. I'd buy a map of a place because I'd taken a trip there or because I had relatives who lived there, something of that sort. And then after a few years I realized that really wasn't the right way to go about it. That's not a collection: it's an assembly of items. And the way I draw the distinction is that selecting a piece for a collection has nothing to do with the individual merits of the item. It's whether there's a potential of relating it to other items. That's what builds a collection: the sum is of greater interest than each of the individual pieces."

And then Mr. Atlas said something that has often occupied my thoughts since. His goal as a collector, he explained, was to tell a "comprehensive story." I confess that this idea escaped—and irritated—me at first. My general impression was that collectors were people with too much time on their hands and too much money in their checking accounts—passive consumers. Collecting had never struck me as a particularly creative act, much less a narrative form. And as a moderately hardworking writer, I was a tad miffed by my host's implication that owning a lot of cool stuff qualified him as some sort of auteur as well.

If Mr. Atlas sensed my skepticism, he did not say so. Nor did he waste time with arguments in support of his thesis. Instead, he set about proving it—with maps. One such demonstration began with a 1718 map of the Mississippi River by the French cartographer Nicolas de Fer. On the inset was a smaller map, showing the entire Gulf Coast region. This rendering of the coast, he explained, was far more accurate than those that had come before it, a big step forward in the cartography of the area. He then showed me other maps bearing that same general image of the coastline, including Guillaume Delisle's *Carte de la*

Louisiane of 1718, "which was widely copied by many other cartographers and was the mother map of that region for half a
century."

Finally, with barely restrained excitement, Mr. Atlas led me
to another part of the house. "Now, look at *this*," he said, carefully taking out one last document. "*This* is where *all of that*
comes from."

On the table in front of me was a work produced not by a
printing press but by hand—what historians call a manuscript
map. Some of the edges had been slightly burned, but I could
easily make out the outline of the Gulf Coast—the same outline I had seen on the other maps. Yes, said Mr. Atlas, I was indeed looking at a one-of-a-kind work, the original map drafted
by the actual person who had surveyed the region nearly three
hundred years ago.

There was something even more haunting about this document than the others I had looked at that day. Looking down at
those uneven lines of ink, it was almost as if I could see the
hand that drew them moving carefully across the page. Then I
began to imagine all the other hands that might have held this
fragile document during its long life: the hardened palms of
sailors and explorers, the sinewy fingers of cartographers, the
pampered mitts of royals, the loving clutches of collectors. Suddenly, I felt a powerful connection to the past—not so much
that I was reliving history but that I was *part* of it, a continent
taking shape right before my eyes.

"This map was done by Mr. Soupart, whom we know
nothing about," Mr. Atlas said. Not even the first name of the

cartographer has survived. Only his map—and only in this house. "It's so rare that you can know the exact source [of a cartographic tradition]. So this was really neat to come across. The other thing is, do you notice that it's burned? Well, we can't prove it but we do know that the ship which in all likelihood was involved in the surveying expedition burned in Pensacola Harbor in 1719. And it wouldn't surprise me a bit if this was one of the things that was rescued."

What a story, I thought. What an absolutely, undeniably marvelous yarn.

Not only did I feel a new admiration for my host but as the afternoon progressed I began to feel something else as well, something I would never have imagined when I walked in the door: I started to think that Mr. Atlas and I had a lot in common. It was not that I suddenly understood his obsession. But the more maps he showed me—relating each one's history, demonstrating its interconnectedness to others—the less his hobby seemed like some arcane and alien pursuit. Was not building a collection, I began to wonder, a little like writing a book? Did not his quest and mine have the same end, the acquisition of knowledge? Did not we each seek to organize this information—one of us on the page, the other on his walls—in such a way that we might, as he had put it, "know the exact source" of a thing? Wasn't a desire to know the source, in fact, precisely what had brought me here?

By the end of the day, I was ready to proclaim Mr. Atlas not only a storyteller but a chronicler of epic events, the Tolstoy of maps. He had devoted his efforts to documenting two sweeping historical narratives. The first one was the exploration of the American West. This period fascinated him in great part, he explained, because the people who did the exploring were often the same ones who made the maps. And since most of those maps were accompanied by diaries, "you can relive the entire trip."

To illustrate his point, he took out a work by Charles Preuss, cartographer for my old pal the Pathfinder. "Here you see all this commentary comes from the Frémont journal," he said, pointing to text within the map. "And some of it is very interesting—like here! 'First view of buffalo, June the thirtieth.' And the campsites are all marked, so you can see how much space they covered on any given day."

His other great passion was the Age of Discovery, a period from which his collection contained a mind-boggling blur of masterpieces. I saw a 1513 work by Martin Waldseemüller, one of the first atlas maps to depict the New World. (It was Waldseemüller who—under the mistaken impression that Amerigo Vespucci had discovered the New World—named the southern part of that landmass America in a 1507 map. He later discovered his error and deleted the name in further editions, but by then the name had stuck.) I saw a rare work from 1534, drafted by Diego Ribero and printed by Giovanni Battista Ramusio, containing the first clearly identified depiction of the Strait of Magellan. I saw the first printed map devoted to the Western Hemisphere, a work by the German humanist Sebastian Münster. I saw maps by Ortelius, Mercator, the French innovator Oronce Finé, the Dutchmen Gerard and Cornelius De Jode, the

Briton John Speed. The list went on and on. Pick a great map from the sixteenth, seventeenth, or eighteenth century, and it was probably in that house.

The one that really seized my imagination, however, was a smaller work by a lesser-known cartographer, Francesco Rosselli. Printed in Florence around 1508, this particular copy of the map had disappeared shortly thereafter. It remained hidden away for hundreds of years—and might never have surfaced, if not for a fortunate turn of fate. "Apparently this copy was discovered during the rebinding of an early book," Mr. Atlas said. "In past centuries, it was not uncommon for other prints, unrelated to the book at all, to be used as filler in the covers: the exterior of the cover would be an animal skin, and then to give it body there would often be paper inside that. And so when the book is being rebound, the conservators know enough to be careful because often there are prints and other things found that way. And that's how this one came to light."

Originally in the hands of a British dealer, the piece was purchased in the 1980s by a San Francisco map seller who, in turn, sold it to a collector in South America. Mr. Atlas became interested in the map when the South American, who "had to come up with lots of money due to a divorce settlement," decided to put it up for auction at Sotheby's.

He knew little about the piece at first. "I had to research it," he explained. "In fact, to me that's really the most interesting part: the work you do prior to making your decision whether to add a piece to the collection or not. And whichever way the decision goes, you've learned a lot. Really, that's what it's about, because it's only a piece of paper, after all. But the key thing is what that piece of paper *represents*. So if you don't know the historical and cultural elements that produced a map, I think you're missing most of the fun."

It turned out that those "historical and cultural elements" were plenty significant. Rosselli's map was the first one to use a projection system that incorporated 360 degrees of longitude and 180 degrees of latitude—making it, "in a very real sense, the first map of the whole world," according to the historian Peter Whitfield. (That "whole world," however, was a quirky one, containing a separate South American continent—labeled "Land of the Holy Cross or New World"—but no North America, only an elongated Asia.) It was also the first map to use the name Antarticus (Latin for "opposite of the north") to describe what was then still a hypothetical southern continent. And it was the first map to incorporate discoveries made by Christopher Columbus during his fourth voyage of 1502–1504.

An important artifact, yes, but more: an extremely rare one. It turned out that only two other copies were known to exist— one at the British National Maritime Museum in Greenwich and the other at the Biblioteca Nazionale Centrale in Florence. Mr. Atlas determined that, if possible, the third one would be his: "It took a lot more digging than usual to find out what this one was. And I said, 'Well, yeah! That's definitely a great one to have. Absolutely!'"

Luck, as it turned out, was on his side. Instead of putting the work up for sale in an auction devoted solely to maps and prints—where it would likely have garnered a great deal of interest—Sotheby's offered it with an assortment of other antiques from the South American collector's holdings, including such unrelated items as furniture. "And it was not pictured in the Sotheby's catalog, either, so, again, attention was not drawn to it," explained Mr. Atlas.

How sweet it must have been to realize this gem was hidden from the view of other potential bidders! Yet how nerve-racking to think they might catch on! I knew enough about the lore of collecting to understand that the most coveted acquisitions are the "finds." There's the story of Henry Stevens, for instance, a prominent bookseller in the nineteenth century who, while visiting the home of an acquaintance, "chanced to notice a small copper globe, a child's plaything, rolling about the floor." Stevens, who later recounted that his host had "picked it up in some town in France for a song," realized that it was no mere knickknack but a world globe from the early 1500s—a priceless, one-of-a-kind artifact now housed at the New York Public Library. By comparison, of course, the reappearance of the Rosselli map was not quite so dramatic. Still, said Mr. Atlas: "I knew there would never be another chance to get it. And there, it's a question of how badly did you want it?"

Just badly enough. Though he refused to say how much he paid for the map, he did note—with discernible glee—that he "got it for a lot less" than the seller had shelled out in the first place. Still, it was clear that the map's monetary value was of relatively little importance to him. "The only time money matters," he insisted, "is at the decision point about whether you're going to buy something."

But if the commercial aspects of collecting weren't what kept him going, then why did he pursue a map like the Rosselli with such fervor? Its historical import? Sure. Its beauty? Of course. Its rarity? Doubtless. The stature it would bring him? Possibly. Yet something else seemed to matter more than all that—something that he never quite expressed but that I nonetheless began to recognize while I watched him pore over that map, his eyes dancing from one place to another as though in search of a previously unnoticed detail. Perhaps it was simple. Perhaps the impulse to collect maps and the impulse to make them arose from the same desire. Perhaps what drove aficionados like Mr. Atlas was precisely what had driven explorers like Columbus: the sheer joy of discovery.

The searching, the dreaming, the navigating of strange seas, the overcoming of obstacles—all leading up to that instant when the unknown is known, the unreachable reached, the unobtainable obtained. In an essay on collecting, the philosopher-critic Walter Benjamin—himself an incurable bibliophile—described that shining moment as "the final thrill, the thrill of acquisition." Columbus had put it another way: the conquest of what appears impossible.

An impossible conquest—I couldn't think of a better description of how Mr. Atlas had come to possess that long-lost document. But I also knew that not all journeys of discovery end so happily. If I needed proof, it was right there in front of me, on that map commemorating the last, lonely journey of history's most famous explorer.

*T*HE ADMIRAL SITS DOWN TO WRITE ANOTHER LETTER. TEN *years have passed and much has changed. He has fallen from favor with the crown, become an object of scorn among his peers. He has arthritis. He has malaria. His eyes bleed, leaving him blind for long periods. At the moment, he finds himself marooned in Jamaica, hundreds of miles from any hope of rescue, his men on the verge of revolt. He writes:* I have not a hair upon me that is not grey, and my body is infirm. *He writes:* I am . . . ruined. *He could have avoided all this, if only he would have retired after his great discovery. ("Almost anyone, it might be thought, would rest content with so much fame, so much wealth, so many discoveries, so dramatic a social rise. But not Columbus," his biographer Felipe Fernández-Armesto would observe. "His sights were always fixed on unmade discoveries, unfinished initiatives, imperfect gains, and frustrated crusades.") The elation of that first triumph was so very sweet but so very short-lived, and he has tried to get it back ever since. Three times he has sailed across the sea, and three times his journeys have ended in frustration. (Once, he was taken back to Europe in chains, due to his misgovernance of the new colonies.) He has made many new discoveries, of course, but they do not satisfy him. He wants more: more land, more status, more fame, more wealth. He writes:* Gold is most excellent. *He writes:* Gold constitutes treasure, and he who possesses it may do what he will in the world. *To seek and to find and to keep on seeking—it's as if his dream has turned into an unquenchable obsession.*

*E*ASING INTO THE SOFA OF WERNER MUENSTER-
berger's midtown Manhattan apartment, I found myself surrounded by unsettlingly beautiful pieces of ancient African art:
brooding figures that lurked in the corners like dim and disqui-
eting memories, phallic totems, wild-eyed masks that stared at
me so intently I almost blushed. They seemed to be carved out
of solid id, those objects, which was only fitting because I had
come here to discuss the shadowy nether regions of the human
mind. Muensterberger, a charming, erudite, and meticulously
attired octogenarian, was a leading expert on the psychology of
collecting. With his bulky glasses, brushed-back white hair, and
wiry eyebrows, he looked every bit like the psychoanalyst and re-
tired professor of psychiatry that he was—and with his Teutonic
accent, the result of a childhood spent in both Germany and the
Netherlands, he somehow sounded the part as well. I had already become fairly familiar with Muensterberger's ideas on collecting through reading his work and interviewing him over the
telephone. What I had not known was that his interest in the
subject was obviously far more than clinical. Yet the fact that he
was himself a confirmed aficionado only made me more anxious to talk to him. If Mr. Atlas had taught me about the beauty
of collecting, I hoped that Dr. Muensterberger could give me
some insight into the darker side of the obsession, a subject he
had explored at length in his book *Collecting: An Unruly Passion.*

Fittingly, we began on the subject of African art. Muensterberger had just visited the home of a fellow collector of such

objects—a man whose life story, he said, offered crucial insights into the mind of the collector. "It happened last night, so it's very fresh in my memory," he explained. "I walked into this house where he keeps his collection, a rather large suburban affair, and hanging on the walls were all these antlers. I said, 'Oh, you collect antlers, as well [as art]?' And he said, 'Well, I was a big hunter.' Then he took me to his basement, where he had an enormous collection of mounted and stuffed gazelles, antelopes, and lions, all of which he had shot himself in East Africa."

Muensterberger explained that the man had never been particularly interested in primitive art until one fateful visit to the western side of the African continent, when, "in a market, he suddenly saw pieces like these." Overcome by the strange beauty of the objects, the man eventually decided to put away his guns and abandon the bloody existence of the safari for the refined world of the auction house. His life had been turned upside down—or so it seemed. But on closer inspection, explained Muensterberger, it hadn't changed much at all: he had simply "switched from the collecting of animal skulls to collecting objects of African sculpture.

"I was sorry that I didn't know him when I was doing my book," he said, "because here is a very essential element of collecting—the hunting. When you talk to collectors, very often aggression, concealed as it may be, plays a role in the obtaining of the object."

Muensterberger believed that, like the hunt for big game, the hunt for rare objects involves what he described in his book as a "search—successful or not—that ever promises hope, suspense, excitement, and even danger." So it came as no surprise to him that his acquaintance demonstrated the same pattern of behavior in the art galleries as he had exhibited in the bush: having set off on an expedition, the man could not rest until he had bagged his prey.

"As a hunter, when he had two or three wonderful animals, he could take off and come back to America. The same is true now that he collects African objects," he said. "He told me he went to Europe to see some dealers in Paris and London and Brussels. Then he showed me an object he had bought. And, you see, when he had found that, he could take it under his arm and come back from the hunt."

Muensterberger felt that such behavior could be traced back to an ancient urge. In our past lives as hunters and gatherers, we humans practiced animism, the belief that when people or animals die, their remaining life force, or soul, is transferred onto ordinary objects. This philosophy, which eased the awful finality of death, has long since been replaced by those that purport to be more rational. Nonetheless, our animistic impulses remain—and, according to Muensterberger, are at the root of our compulsion to collect. As he put it in the book, "There is reason to believe that the true source of the habit is the emotional state leading to a more or less perpetual attempt to surround oneself with magically potent objects. . . . The compelling concern [is] to go in search, to discover, to add to one's store, or holding, or harem."

I asked him why the sense of discovery seemed to play such an important role for the collector. "Think of the word: *dis-*

cover—to take the cover off and see what's there," he told me. "It goes very deep for the collector: I want to find out. And what you really want to find is, Where do I come from? What is the source? That is discovery—finding something no one knew before, and you didn't know before."

That jibed with what I had learned from Mr. Atlas (whose pursuit of the long-hidden Rosselli map had begun—quite literally—with someone "taking off the cover"). But Muensterberger believed that such discoveries are not without a downside. While they can bring the collector much happiness, he noted, the elation is usually short-lived, giving way to renewed feelings of frustration. And so the hunt goes on. "The quest is never-ending," he wrote in his book. "It is, as one can see time and again, bound to repeat itself, while the ultimate pleasure always remains a mirage."

Most collectors seem to accept, even embrace, this contradiction—aware that for the aficionado, as for the adventurer, the journey is better than the destination. Mr. Atlas, for instance, seemed to have his habit well under control, buying only a few new maps each year at this point in his life. ("In fact," he had told me, "one of the reasons that I've gotten involved in exhibits and the lecture series and other aspects of maps is that . . . I didn't want my interest to be dependent on the next acquisition.")

But for a few aficionados what begins as harmless hobby can devolve into what Muensterberger described as "an all-consuming passion, not unlike the dedication of a compulsive gambler to the gaming table—to the point where it can affect a person's life and become *the* paramount concern . . . overshadowing all else: work, family, social obligations and responsibilities. We know of numerous cases in which moral standards, legal considerations, and societal taboos have been disregarded in the passion to collect."

History has seen more than a few out-of-control collectors, such as the notorious Don Vincente, a bookseller and former monk whose malignant bibliomania led him to commit no fewer than eight murders in nineteenth-century Barcelona. But while it is tempting to dismiss such zealots as a mentally disturbed minority, Muensterberger warned that even a generally well-adjusted and honest person can be tempted to transgress laws or moral codes—in part because overcoming the obstacles that lie between the collector and the desired object is exactly what makes the experience so rewarding. "Look, I spoke of the discovery," he said. "*Découvrir* in French, *entdecken* in German—I could use any number of languages, and it has more or less the same meaning: to take off the cover. So, as a collector, what you do is look for something that you're not supposed to see or possess. And so, intrinsically, whatever you discover or obtain by discovery is taboo in one way or another.

"There is always the chance of having trespassed," he added, glancing slowly around the room at his beloved possessions. "Even with these African objects it's true."

He had been collecting such pieces for no less than seven decades: "I discovered African art the first time consciously when I was eight years old in the house of a distant relative. I was so taken by it that, when I was thirteen, I went to the flea market in Amsterdam and saw a piece. The man said, 'It comes from the Dutch colonies—Indonesia.' I said, 'No, it comes from Africa.' He said, 'No, stupid boy.' But it clearly was a very nice African horse and rider. That was the first piece I ever bought."

In the years since then, he had tried to avoid buying items that are stolen from—or are sacred to—the people whose ancestors made them, instead purchasing only objects that either are meaningless to them or have been abandoned.

"But some of the objects have been discovered only recently, buried underground. Now, to whom do they belong? In Italy, anything in the ground belongs automatically to the state. In Africa, they have laws—but there are tribal laws and national laws and provincial laws," he said with a sigh.

"It's all very muddy," he added. And then, for a second, the old man's gaze seemed to lose itself on one of the sculptures. I have no idea where his thoughts had taken him at that moment. My own had drifted, as they did so often in those days, back to Gilbert Bland.

OF THE MANY WONDERFUL YARNS SURROUNDING THE LIFE of Christopher Columbus, my favorite is the story of the Unknown Pilot. The tale is said to have taken place around 1480, when Columbus lived on Porto Santo, one of the Madeira Islands off the North African coast. It begins when a crippled vessel, floating at the mercy of the waves, lands on the island. All of the ship's sailors have perished, except for the helmsman, who is on death's doorstep himself. Seeking to nurse the mariner back to health, Columbus takes him to his house. There, the ailing man reveals an astonishing secret: another world exists on the far side of the Ocean Sea (as the Atlantic was then known). The man has been to these distant shores himself, after his ship blew off course during a storm. With the pilot's help Columbus crafts a map showing the route to these new lands. Then the man dies, leaving our hero alone with the secret that will change the course of history.

The story of the Unknown Pilot is a myth. We can be fairly sure of this not only because it lacks supporting evidence but because it *sounds* so much like a myth. Substitute the Unknown Pilot for the Lady of the Lake and the secret map for Excalibur, and suddenly we're talking about not Christopher Columbus and the New World but King Arthur and the Holy Grail. In *The Hero with a Thousand Faces,* his classic study of legends, Joseph Campbell wrote that a key starting point for the hero's journey is "a protective figure (often a little old crone or old man) who provides the adventurer with amulets against the dragon forces he is about to pass. . . . What such a figure represents is the benign, protecting power of destiny. The fantasy is a reassurance—a promise that the peace of Paradise, which was known first within the mother womb, is not to be lost."

If, as has often been said, myths teach us fundamental truths about ourselves, then the Unknown Pilot legend demonstrates the powerful spell maps cast on the human imagination. Throughout the centuries people have viewed maps not just as useful navigational tools but as enchanted objects—what Campbell called "amulets" and Muensterberger described as "power-imbued fetish[es]." Columbus himself, for instance, seemed to think maps were endowed with a force that transcended mere matters of geography. They stoked his imagination, inspired the flights of fancy that made his great discovery possible. The sixteenth-century chronicler Bartolomé de Las Casas wrote that when Columbus gained access to a map from the Florentine scientist Paolo del Pozzo Toscanelli it "set [his] mind ablaze." Likewise, Ferdinand Columbus reported that after another acquisition of sea charts, his father's passion was "still more inflamed." A cartographer and dealer himself, Columbus drew maps, dreamed maps, meditated upon maps, listened to maps, acquired them at every opportunity (sometimes under very questionable circumstances). Even in 1492, after sailing into the unknown, far past the point where maps would be of any practical use to him, he continued to pore over them, fingering their edges the way a child might rub the satin lining of a blanket, as if their mere presence could ward off danger and make land appear on the horizon. I have no hesitation about labeling Christopher Columbus a cartomaniac.

In the centuries that followed his discovery, maps became ever more scientific and less whimsical. Even so, people continued to be drawn to them for reasons that had nothing to do with utility. In 1570, for example, John Dee, the Elizabethan mathematician and mystic, wrote this about why people collect maps:

Some, to beautify their Halls, Parlors, Chambers, Galeries, Studies, or Libraries with, some other for their

own journeys directing into far lands, or to understand
other men's travels, liketh, loveth, getteth, and useth,
Maps, Charts and Geographical Globes . . .

But what is behind the liketh and the loveth, let alone the
getteth and the useth? What is the specific allure of maps, especially old maps? Why are some people drawn to maps instead
of, say, books, porcelain, beer cans, dolls, or any of a million
other objects they might collect?

When I put such questions to Muensterberger, he said the
map lovers he has met tend to come from broken homes or
families that have moved around a good deal. They throw
themselves into their hobby, at least in part, as a way to connect
with a parent or to ground themselves in a more permanent
sense of place. "Looking for maps, especially antique maps," he
told me, "is really looking for the past—Where do I come from?
Who were my ancestors?—and, symbolically, finding security."

Harriette Kaley, a Manhattan clinical psychologist and psychoanalyst who has also studied the phenomenon of collecting,
offered a similar view. "Analysts refer to a person's early childhood as his or her prehistory," said Kaley, herself a minor map
enthusiast. "And in fact, people remember that period of their
lives as a kind of fantasy, because an adult is unable to recapture
the way his or her thought process worked as a child. For each
of us, our early life seems like the distant past, and in that sense
it's like an ancient land—far-off, foreign, and unknown. It occurred to me that it's not unlike the way fairy tales begin: 'Once
upon a time in a faraway place.' And I think there must be

something in map collecting that taps into that. In some sense, old maps reach back into a part of life that you can't quite grasp, and give you a sense of where you've come from. They give you a feeling of being rooted."

Does any of this help explain the actions of our Antihero with a Thousand Faces, Gilbert Bland? His critics would probably say no. Most of those who came in contact with him told me that, in retrospect, they believed his interest in maps was never sincere; his only real passion was for a good scam. This view was somewhat bolstered by Bland himself, who reportedly told the FBI he had happened upon maps by accident. Nonetheless, I tend to agree with Joseph Campbell: "As Freud has shown, blunders are not the merest chance. They are the result of suppressed desires and conflicts. They are ripples on the surface of life, produced by unsuspected springs. And these may be very deep—as deep as the soul itself. The blunder may amount to the opening of a destiny."

Of all the criminal enterprises a would-be felon might select, I can't believe Bland chose map theft entirely on its merits as a racket. I won't presume to guess what "suppressed desires and conflicts" might have drawn Bland to maps, other than to note that his somewhat rootless childhood seems to fit a general pattern for cartographic connoisseurs as outlined by Muensterberger. And while he was not, strictly speaking, a collector, he did seem to exhibit the behavior of an obsessive aficionado. When I look through the entries of his notebook, for instance, I see someone who operates with the single-minded determination of a hunter. Each new map is like an elusive quarry, meticulously stalked down, first on the Internet and then in the library itself. The hunt is full of danger (*Can't these people leave? I can't do it now. OK now*), but it is this very risk, perhaps, that makes the eventual conquest so exhilarating (*These 2 are done now. Thank God!*). And by the end the thrill seems to have become addictive—so addictive that it's almost as if one's whole life hinges on the next conquest (*The Bowen Atlas—of all the bad luck—What is going on here. Am I not going to get these Bowens? What [will] become of me?*). Which brings us back to the tale of the Unknown Pilot. It turns out that there's a second version of the legend, one that offers a different lesson about the allure of maps. In this one the sailor does not die of natural causes. In this one Christopher Columbus bullies the man into supplying him with the map—then kills him to keep the secret all to himself. In this one the great discoverer becomes a common criminal.

ONCE DABBLED IN A SORT OF MAP COLLECTING MYself. When I was twenty-five years old I covered the bedroom walls of my Chicago apartment with street maps of every place I had ever called home, a less than cosmopolitan list that, in addition to the Windy City, included the Illinois burgs of Downers Grove, Carbondale, Champaign-Urbana, and Springfield, as well as Washington, D.C., and Manchester, New Hampshire. I can't say for certain what prompted me to do this. Perhaps I was simply looking back at my childhood through those maps, aware that, with my father in failing health, the full weight of adult life's realities and responsibilities would soon be upon me. They were still on the walls when he died, but fate soon put an end to the collection. An arsonist set fire to the building, leaving the apartment in ruins. (I was lucky—in more ways than one—to be spending the night elsewhere.) I learned a lot from that blaze. It taught me that I could do without my possessions just fine, but I could not do without my friends and family. It also reminded me to look forward instead of dwelling on the past. I never retrieved those maps from the wreckage but left them hanging there, clouded with soot, like so many faded memories.

Nor was I seriously tempted, in all the time I spent working on this book, to purchase an antique map. Having reproductions of them in books was enough for me; unlike Mr. Atlas, I did not need direct contact. The truth is, I have never done much collecting at all. A few baseball cards as a kid, some 1950s-era slide projectors as an adult—but even these were half-hearted pursuits, inspired more by boredom than by any overriding passion. I always told myself that I was free from the compulsion.

I was wrong. That fact hit me like a bus during the afternoon I spent sitting on Werner Muensterberger's sofa in Manhattan. The elderly psychoanalyst was holding forth on the relationship between collecting and hunting when it suddenly occurred to me that, without really noticing it, I had devoted years of my life to building a collection, and a very extensive one at that. It was sitting on the shelves of my office back in Chicago, neatly displayed in a growing series of three-hole binders. Gilbert Bland's birth, marriage, and divorce records. Gilbert Bland's military documents. Gilbert Bland's criminal records. Transcripts from his trials. Gilbert Bland's prison correspondence. Documents related to his business and real estate dealings. Forms from his wife's bankruptcy case. A copy of the notebook Gilbert Bland left at the Peabody Library. Photographs of the maps he stole. Hundreds of pages of testimony from those who came in contact with him. I had never spent so much time, effort, or money acquiring any group of objects in my life.

All journalists keep records—but these scraps of paper had become much more than that. I realized that they were my amulets, my fetishes. Every time some new bit of Blandabilia arrived in the mail or turned up in an archive, I would rush to examine it with a feeling of giddy anticipation, sure that it would fill in some key corner of the map I was trying to create. Often it did just that—but invariably it would also raise new questions, so that as one blank space was filled another would open up, my map becoming simultaneously larger and smaller, Bland both more and less real. Excitement would then give way to disappointment—but just as quickly I would think of some new document or witness that, if tracked down, might provide just the right piece for the puzzle. And so the process would begin again, and I would set out in search of yet another discovery.

But in "taking off the cover," I sometimes saw more than I wanted to see. Prying into other people's lives is a necessary, if unpleasant, part of a reporter's job. In this case, however, it seemed even more disagreeable than usual. Not that I felt particularly sorry for Bland. His own actions had transformed him into a public figure, and, whether he liked it or not, public figures merit public scrutiny. Yet, because of a long-ago tragedy in my own family, I knew that the map thief would not be the only one to suffer. During the Great Depression my grandfather was sent to prison for bank embezzlement—a traumatizing event that left lasting scars on my mother's psyche and indirectly shaped my own personality. This incident, still a touchy subject in my family, assumed an increasingly central place in my thoughts as the investigation progressed.

Perhaps because of this, I felt particularly uncomfortable delving into matters that affected Bland's wife and children. When, for instance, one of Bland's daughters by his first marriage agreed to speak with me after much hesitation, the conversation turned out to be an awkward one for both of us. Heather Bland knew very little about her father, and the few memories she did recount were mostly unpleasant. I ended up learning more about her pain than I did about his life—and, though I was grateful to her, I was also relieved when our conversation concluded. I learned a similar lesson once in Florida. I was talking to people who lived on Bland's street, trying to add one more precious little piece to my collection, when a neighbor stopped me and pointed to a group of children playing with skateboards at the end of the block. "One of those kids is his," the man said sternly. "Is it really so important for you to be here right now?" I thought about it for a second, shook my head no, then got into my car and drove off.

Yet as unpleasant as such incidents could be, they were not the creepiest aspect of my pursuit. I simply cannot explain how surreal it was to spend years tracking down a man I'd never met, how disconcerting it felt to know someone so well and yet not know him at all. At first, Bland had seemed to me an exotic and intriguing figure—but, as is often the case with familiarity, the more I learned about him, the less interesting he became. He was, I ultimately determined, a fairly unexceptional person who had happened to commit a fascinating crime. At some point I realized he was not even someone I would normally want to know. But even that did not stop me from seeking him out. I had to face the fact that my search had become a compulsion—not so different from the one that lured Bland into those rare books rooms, perhaps, or the one that Rudyard Kipling described in his 1898 poem "The Explorer":

> "Something hidden. Go and find it. Go and look
>   behind the Ranges—
> "Something lost behind the Ranges. Lost and waiting
>   for you. Go!"

Then one day I came to an even more disturbing realization—that my identity and the map thief's were somehow starting to converge, that he had taken up permanent residence on the edges of my consciousness, a cipher on which I was projecting musings and misgivings and fears that had nothing to do with the case. And at that moment I understood that I was no longer searching only for the map thief. My quest had gone beyond Bland: I was now hunting down some enigmatic citizen of my own psyche, a persona I did not know and did not particularly look forward to meeting. There seemed to be no going back, however. Remembering what Aldous Huxley had written about what lurks in "the antipodes of the mind," I sensed that I was on a collision course with one of those "strange psychological creatures leading an autonomous existence according to the law of their own being." It was one discovery, I feared, that would bring no joy at all.

DETAIL FROM HERMAN MOLL'S 1719
MAP OF NORTH AMERICA, DEPICTING
CALIFORNIA AS AN ISLAND.

CHAPTER 11

The Island of Lost Maps

*T*HE HISTORY OF CARTOGRAPHY IS FULL OF PE-
culiar islands. One of the items Gilbert Bland stole from the
University of Virginia, for example, is the cartographer Herman Moll's eighteenth-century map of North America. It
shows a continent that looks a lot like the one we now inhabit,
except for one striking detail. Running parallel to the West
Coast all the way from what is now Canada to Mexico is a
sprawling independent landmass—a famous and widely repeated cartographic miscalculation known as the Island of
California. Other charts contain isles even more bizarre. A
twelfth-century map by the Arab geographer al-Idrīsī shows El
Wakwak, an island said to be filled with trees whose fruits,
shaped like the heads of women, continually cry out the meaningless chant "Wak-Wak." The famous Hereford *mappa mundi*,
drawn around 1300, has an isle on which "sirens abound," ac-

cording to the accompanying text, while another world map of the period contains an island inhabited by "men who murdered their fathers." The Catalan Atlas, made around 1375, shows one isle that "produces all crops and all fruits without any need to sow or plant" and a second where "there are trees that yield birds as other trees yield fruits." Martin Behaim's history-making globe of 1492 shows the Islands of the Satyrs, whose residents "have tails like animals," as well as a pair of Indian Ocean landmarks named Masculina and Feminea. "One of these islands [is] inhabited by men only, the other by women only, who [meet] once a year," reads the legend. Even the great Ortelius published a map in 1570 that depicts several completely mythical islands in the North Atlantic, including Drogeo, purported to be inhabited by cannibals; Icaria, whose king was said to be a direct descendant of Daedalus; and Saint Brendan, where the legendary Irishman is said to have brought a dead giant back to life.

From the earliest of times islands have had a special place in the human imagination. So close to our own world yet so out of reach, they have been the landscapes where no life-form was unimaginable, no occurrence impossible. Homer spattered *The Odyssey*'s oceans with isles such as Aeaea, where the sorceress Circe transformed men into pigs; Cyclopes Island, where one-eyed giants dined on human flesh; and Lotus-land, where visitors, having consumed the local flora, forgot their friends and lost all desire to return to their native lands. Other classical writers described islands on which those favored by the gods lived untouched by sorrow: Hesiod and Pindar called these the Isles of the Blessed, while Plato provided the first written account of Atlantis, the splendrous island that disappeared into the ocean. Celtic myth was full of enchanted outcroppings—islands of laughter, islands "full of men agrieving and lamenting," islands of women, islands shared by the living and the dead, islands where time stood still, and islands that could be seen only by the elect, rising out of the sea for one adventurer and vanishing back into the depths for another. Arthurian legend told of Avalon, a paradisal island inhabited by nine sorceresses, where the great king was taken after being mortally wounded. The *Arabian Nights* described one island that was home to "a bird of monstrous size called the roc, which fed its young on elephants," and another that was inhabited by the Old Man of the Sea, a huge beast that enslaved visitors. The Tantric books of medieval and modern India told of an Island of Jewels, where a goddess lived in a grove of wish-fulfilling trees. Thomas More put Utopia fifteen miles from the coast of the New World. Shakespeare set *The Tempest* on an island "full of noises, sounds and sweet airs," ruled by a noble magician. Native American island legends included the tale of Maushope, a giant on Martha's Vineyard, who pulled whales from the sea with his bare hands and ate them raw. J. M. Barrie set *Peter Pan* on the Never Land, an isle inhabited by lost children, "where all the four seasons may pass while you are filling a jug at the well." The film director Merian C. Cooper dreamed up an island "way west of Sumatra," then populated it with huge lizards and a petulant primate named King Kong. And while such enchanted islands no longer can be found on maps, they continue to dot the dark seas of our collective unconscious. It was no coincidence, for instance, that, when Michael Crichton needed a setting for his modern-day dinosaur epic *Jurassic Park,* he chose a volcanic seamount whose "forested slopes were wreathed in fog, giving the island a mysterious appearance."

Now let me tell you about another wondrous isle, one I saw with my own eyes. Like some evanescent island of legend, this one was elusive. Had you been with me when I first beheld it, you might not have noticed an island at all—just a mound of plastic bags, file folders, a zip-up art portfolio, and a U-Haul moving box, sitting in the middle of a vast conference table at the FBI office in Richmond, Virginia. In my eyes, however, it was a strange, sad world unto itself, full of marvels. All its inhabitants were exiles: Indians, Eskimos, cossacks, conquistadors, emperors, presidents, chieftains, queens, kings, soldiers, sailors, monsters, mermaids, serpents, seals, whales, lions, beavers, elephants, polar bears, fire-breathing horses, gods, and angels. As I approached its shores, they all seemed to be speaking to me at once, telling a million different stories in a thousand different tongues—tales of oceans crossed and shorelines glimpsed, of new worlds explored and old orthodoxies exploded, of empires gained and lost, of wars waged and genocides committed and peoples enslaved, of forests felled and cities built and borders drawn and railways laid and prairies cleared. They were chanting the whole history of the last five hundred years, the inhabitants of that tiny island, but they were also reciting another tale, one about how history can be untold with a few silent strokes of a razor blade. I had been seeking this place almost since my journey began. Now here it was, before me at last: the Island of Lost Maps.

The lord of this curious domain was seated across the way. Gray Hill was a lanky, middle-aged man who towered over his island the way Gulliver overshadowed Lilliput in Jonathan Swift's *Travels into Several Remote Nations of the World.* I knew that Hill had not come to these shores by choice. Like Gulliver and the heroes of a thousand other island tales, he was a castaway, brought here by the improbable tides of fate. Hill was an FBI agent, not a map collector—but, in the famous words of a shipwrecked traveler from *The Tempest,* "misery acquaints a man with strange bed-fellows." It was now July 1996, and he had already been stranded for some six months. He did not know it yet, but he was destined to remain for years, as if the Island of Lost Maps was enchanted and Hill was being held under the spell of a sorcerer. And, in a certain sense, that was exactly what was happening.

IN HIS BOOK ABOUT COLLECTING, WERNER MUENSTERberger wrote that the dealer of antiquities is sometimes endowed with almost magical powers. Many aficionados, he observed, view the materials they collect not as inanimate objects but as fetishes with "a soul or a life-force of their own." As a result, wrote Muensterberger, "the relationship between collector and dealer is different from any customary buyer-seller contact . . . largely due to the intrinsic power that accrues to the dispenser of magic provisions. This predicament is one of the most potent assets of the successful dealer, whose role is often closely akin to that of physician, priest, or shaman."

For those who had done business with Gilbert Bland, the mysterious dealer from South Florida must have indeed seemed like something of a shaman at times, able to conjure up maps as fast as he could snap his fingers. *Abracadabra, yield us an Ortelius! Open sesame, here comes a Jefferys!* One of those who got a firsthand demonstration of Bland's black art was a prominent Florida collector who visited the map thief's booth at the Miami International Map Fair in 1995. He came away—happy but perplexed—with two copies of a hard-to-find Florida map from 1845. "He had three of them," the collector said of Bland's inventory. "Those were the first three I'd seen [for sale] in six years."

To possess what others covet—what a heady feeling that must have been for the map thief, especially after his frustrations in the computer business. Systems integration, RAM upgrades, peripherals—people may need these things, but they do not hunger after them, do not stay awake at night aching for their feel, their look, their smell. Even before his computer firm went belly-up, Bland must have known that he was trafficking in mundane essentials—the same ones everybody else was trying to hawk during the 1990s. But to be the bearer of rare maps—well, that changed everything. Suddenly Bland was golden. Suddenly he was in demand. Suddenly he was a genie with the power to grant—or deny—people's fondest wishes.

I don't know how Bland felt about his newfound power, whether it thrilled him or scared him or gave him a good laugh.

What I'm quite sure about, however, is that he knew how to wield it. Even after he was arrested, his magic did not fail him. In a strange way, it became even more potent. Bland had been in legal trouble before, and he was skillful at working the courts. He knew that the judicial system was a lot like the antiques business: if you had something that the other side desperately wanted, you were likely to get a very favorable deal. And so, before surrendering to police, Voodoo-man Bland pulled off his most remarkable piece of prestidigitation: he made hundreds of old maps disappear, as if into thin air. Somewhere he was holding New Jersey, Virginia, and Maryland, as well as Italy, Sweden, and Norway. He had the fortifications of Montreal. He had the Missouri Territory. He had the Empire of China, the Empire of Japan, and India beyond the Ganges. He had the Eastern Hemisphere, the Western Hemisphere, and the North Pole. He even had the trade winds locked up somewhere.

LIVE IN FEAR OF GETTING THESE THINGS WET."

As he carefully unfolded one fragile map after another, Special Agent Hill cast a suspicious glance at the can of soda in my hand. Spreading out on the table before us was a haphazard collection of cartographic gems. There were maps by Ortelius, Ogilby, and the Pathfinder. There were Hondius and Mercator. There were Jacques Le Moyne de Morgues, who visited Florida in 1564 and produced the first detailed map of the region; Guillaume Delisle and Jacques Nicolas Bellin, two leading French cartographers of the early eighteenth century; Thomas Jefferys, the official geographer of Britain's King George III and creator of one of the first important atlases of North America; and many others, including the great explorer James Cook. Glancing from one old map to another, I was once again struck by their strange beauty, a kinetic clash between austere grids and voluptuous contours, empirical reality and romantic fantasy, as if the cartographers could never decide whether to make rigorous scientific diagrams or erotic doodlings of Mother Earth. It was disconcerting to see them heaped up like that, centuries of mesmerizing and historically crucial documents scattered unceremoniously across the tabletop like a bunch of wallpaper samples. And, had the thief not been caught, that would have been their fate—decorating walls. Now it was Hill's job to undo Gilbert Bland's crime spree page by page, returning each of these orphaned masterworks to its proper home.

Hill pulled out a map from the 1607 edition of the famous Hondius-Mercator atlas. "This is another one that I don't have any idea where it came from," he said glumly. Then he turned to me, his frown slowly transforming into a conspiratorial grin.

"Hey," he said, holding out the map with a wink, "you got your checkbook with you?"

Hill kept surprising me. Prolonged childhood exposure to the TV series *The F.B.I.* had led me to sort of expect that all G-men would be like the stone-faced Efrem Zimbalist, Jr., who played his role as though possession of a personality could land you on the Ten Most Wanted List. But the amiable Hill, wearing a comfortable blue blazer as he reclined beneath a portrait of grim old J. Edgar Hoover (who looked like he had laced up his girdle a little too tight for the photo shoot), seemed to be a different sort of agent entirely. I had no doubt he could be a tough guy when the situation demanded, but at the moment Hill's most menacing quality was his booming tommy gun of a laugh.

Then again, everything seemed a little topsy-turvy on the Island of Lost Maps. Normally, Special Agent Hill's job was to track down lawbreakers, but now his role had been reversed: he hunted victims. This offbeat assignment had fallen into his lap shortly after Bland's arrest, when the map thief was extradited to Virginia. Federal prosecutors in Charlottesville—Hill's base of operations—indicted Bland on charges of stealing objects of significant cultural heritage and transporting stolen goods across state lines. If convicted on those charges, the map thief would have faced up to twenty years in prison without the possibility of parole and a fine of half a million dollars. Moreover, state courts in North Carolina and Delaware would soon file additional counts against Bland—and other states were threatening to follow. In short, the map thief was in what some old cartographer might have called an *orbis terrarum* of trouble. On the other hand, he had what one of Hill's associates later described as "a very effective bargaining chip."

Not only was Bland the only one who knew where the maps were hidden but he was also the only one who knew where they all came from. As a result, explained Hill, the feds faced a tough dilemma: "Do we take our pound of flesh, or do we say to the defendant, 'Okay, fine, we'll take eight ounces of flesh in return for information about what you had access to and what you have stolen'? We decided it was worth getting eight ounces of flesh, and maybe even a little bit less, if we could get all the maps returned."

Bland finally agreed to a plea bargain in which he would receive a reduced sentence and limited immunity from further prosecution. In return, he promised, among other things, to cooperate fully with FBI agents in their attempts to get the maps back to their rightful owners. So in February 1996, while jailed without bond, the map thief directed FBI agents to a storage locker in Palm Beach Gardens, Florida, that he had rented under an assumed name. When they looked behind its bright orange doors, they discovered an extraordinary booty: some 150 maps in all, a few of them more than four centuries old. Taken together with the 100 or so other maps that authorities would gather from Bland's clients across the country, the thief's total collection of some 250 maps had a market value estimated at as much as half a million dollars.

Federal authorities promptly held a self-congratulatory press conference to announce the maps had been "recovered, unharmed." But recovering them was one thing; returning them proved to be something else entirely. My visit came a full five months after the FBI had obtained the maps. Yet despite an exhaustive search, Hill had been able to positively identify the owners of only about seventy—less than one third of the total.

His efforts had been slowed by three big obstacles, the first having to do with the documents themselves. As a rule, maps are unmarked; many lack even a page number. Some institutions do, in fact, put stamps or other types of identification on their maps, but to many librarians this practice is repugnant, the equivalent of stenciling PROPERTY OF THE LOUVRE across the Mona Lisa. Nor do libraries always keep inventories of the maps that are bound in books—so even if they discover one missing, they can rarely be sure when it disappeared. As a result, FBI technicians were being forced to match each stolen map, jigsaw-puzzle style, with each damaged book, using ultra-violet-light technology to make sure the edges lined up precisely and the paper stock was exactly the same on both sides of the cut. It was a labor-intensive and time-consuming process—and, for all that, it did not always produce the desired results. "Just because someone is missing a map does not necessarily mean that the map I have matches that book," explained Hill. "I have duplicates—four or five of the same map. Some of the matches look good to the naked eye. But under the ultraviolet light, it's a different story."

A second problem was the attitude of certain librarians. Although the vast majority of institutions cooperated willingly with his investigation, Hill was surprised to find resistance from some quarters. A couple of libraries, for instance, had rebuffed his initial inquiries, asserting their records were private and Hill would need a subpoena to see them. "We explained to them, 'It's not *our* maps that are missing,'" Hill said, "'And any information you have concerning the theft would be directly beneficial to you as well as to us.'"

Even so, at least one institution had chosen to remain in denial about the thefts, even when the evidence was incontrovertible. "I talked to one librarian who said, 'There's no way he could have stolen anything out of here.' Well, I said, 'I just know one thing. I know that Mr. Bland told me that he came to your library and stole maps.' But they won't accept it. They will not believe that they have had anything stolen."

Or maybe they believed it all too well. As Robert Karrow, curator of maps at the Newberry Library in Chicago, once told me: "A lot of library thefts have gone unreported in the past. You know—you're embarrassed, and maybe you say to yourself, 'What will the donors think?' And you're reluctant to talk about the whole issue because you don't want to give the crazies ideas."

Hill fretted that a similar rationale might now be keeping other victims from coming forward. But he knew that there was absolutely nothing he could do about it, except throw up his hands and say, "If you sit there and tell me that your security is so great that Bland can't possibly have stolen anything from you, well then, I have nothing to return to you."

The agent's other big frustration was Bland himself. Under the terms of his plea bargain, the map thief was required to provide full information about what libraries he had visited, what maps he had taken, and what had become of those items. To a great extent, he had done just that—but Hill knew that Bland was not being completely truthful. The thief was insisting, for example, that many items in the FBI's collection were rightfully his, having been legally obtained from other dealers. He could not provide documentation of the alleged purchases, however, and when Hill set a little trap for him—showing him a couple of items that FBI technicians already knew to be the property of a certain library—the map thief fell right into it. "Those are mine!" he reportedly declared.

"I have never seen anything proving that he had *any* legitimate purchases," sneered Hill. "Why buy something if you can steal it? It gives you a much higher margin of profit."

True, but if all those unaccounted-for maps weren't Bland's, whose were they? For now, Gray Hill had no answers. "Some universities," he said, glancing at the pile with a look of resignation, "may not even know they are missing these items."

ROBINSON CRUSOE ON THE ISLAND OF DESPAIR, Napoleon on Elba, Odysseus on Ogygia, Scarface Capone on Alcatraz, the souls of the dead on Dante's Purgatory—in literature, myth, and history alike, islands have frequently been places of waiting. And thus it was on the Island of Lost Maps. Time moved so slowly there that Special Agent Hill might have been forgiven if he, like Crusoe, sometimes felt that "all possibility of deliverance . . . seemed to be entirely taken from me."

It was not the maps that wore on his nerves. He found the history of cartography "an interesting field," about which he had become something of an accidental expert—able, for instance, to quickly spot the difference between almost identical maps. Nonetheless, Hill's job had become such a bureaucratic nightmare that it was starting to look like he would still be marooned on these shores long after Gilbert Bland got out of prison. "I'm getting to know my stock very well, going through it time and time again," he said with a sigh, placing another centuries-old piece of paper atop the heap.

In recent weeks the island had seen a number of visitors. Like adventurers in search of lost treasure, librarians from all over the country had come bearing slashed-up books. "Some people leave very happy," Hill reported, "and a few leave very upset."

At least one library got more maps from Hill than staff members had even known they were missing. But relatively few of the visitors were so lucky. Another institution was looking for "probably thirty or forty maps," all of which, in theory at least, corresponded with those in the FBI's collection. "Well, they came down and brought their books, and we started trying to match up maps," he said. "We matched up *nothing*. The books are missing these items—but they just didn't match up with the ones I have."

Despite such setbacks, Hill was not giving up. "It becomes an obsession," he told me, "to try to get rid of as many of these things as possible." His next step was to compile a volume containing photographs and precise descriptions of all the lost maps. Copies of this "big black book," as some librarians would later refer to it, were to be circulated all over the country. "Before, the way we were going about it was, 'You tell me what you're missing and we'll see if we can match it up,' " Hill explained. "Now we're going to go back out and say, 'Hey, we have these items; you look and tell us whether you're missing any of them.' "

Hill's persistence would eventually lead to a number of successes. Six months after my visit to Richmond, the British Columbia Archives in Victoria discovered that a certain James Perry had visited in October 1995, departing with twenty maps. Bland had not mentioned this episode to Hill. A similar stop at the University of British Columbia in Vancouver had also apparently slipped the thief's mind. With the help of Gray Hill's big black book, these two institutions would get back many of their maps. And, as the months wore on, Hill would find homes for dozens of other misbegotten documents. Even so, many mysteries would continue to hang over the Island of Lost Maps.

THERE'S A FAMOUS PASSAGE FROM *ROBINSON CRUSOE* in which the shipwrecked narrator discovers a single, inexplicable footprint in the sand. Having been alone for many years, he is instantly overcome by "wild ideas" and "strange unaccountable whimsies." At one point he concludes that cannibals have invaded his island, only to decide that, no, he himself left the print. Other times he "fancy'd it must be the devil . . . for how should any other thing in human shape come into this place?" Such "cogitations, apprehensions, and reflections" haunt him for "many hours, days, nay, I may say weeks and months."

It is no exaggeration to say that, like Crusoe, I had begun to experience "innumerable fluttering thoughts, like a man perfectly confused and out of my self." Increasingly, I felt as though I were chasing a ghost—traces of his presence all over, but he himself nowhere to be found. And in the months to come that ghost would begin to feel more and more like my own shadow, until it could sometimes seem as if Gilbert Bland was on my trail and not the other way around. But now, as I watched Gray Hill ponder one map after another, meditating on the mystery of each one's origin, I realized that at least I was not alone: the agent and I were following the same footprints. This was a strangely comforting thought, yet I knew that it was nothing he and I could discuss. Not only did the rules of our respective professions prevent us from comparing notes but Hill obviously had little time for small talk. He had more pressing cases to contend with, including the recent and widely publicized murders of two female hikers in Shenandoah National Park. And, to tell the truth, I was in a bit of a rush myself. At long last I hoped that those footprints might lead me face-to-face with the one who had left them.

CHAPTER 12

Eldorado

HEN CAME THE HOUR WHEN I DISSOLVED INTO the map. Eighty miles an hour in a nice new rental car, and there I was, drinking my Dr Pepper and tapping my foot to something random on the radio, when all of a sudden—wooosh!—I'm hurtling backward through time and forward through space all at once. Part of me was still on Interstate 85, south of the Virginia–North Carolina state line, but part of me was on another road, a ghost road, the one that lay along this route before the concrete and cars, the horses, the white people. It was called the Great Indian Trading Path, one of the oldest travel routes in America, a byway that once snaked for more than five hundred miles from Virginia to Georgia. I had seen it on an old map of Carolina—engraved by James Moxon and published by John Ogilby in 1672—one of the same maps, in fact, that was in Special Agent Gray Hill's collection. The trad-

DETAIL FROM JOHN OGILBY-JAMES
MOXON'S CAROLINA MAP OF 1672.

ing path was only a thin gray line on that document, so sketchy that I had to squint to make it out, but now it seemed to be reaching out to me like a runway to an airplane, so real and so close that I could almost taste its hot dusty air. So this is what happens, I thought. This is what comes of so many hours spent peering over the edges of old maps. Sooner or later you fall in.

Or maybe it's just what comes of being tired and hungry and anxious about a meeting that may or may not take place an hour down the road. But whatever was causing this madness, I did not fight it. Give yourself over to the map, I told myself. March down that trading path. Scale those peaks—the ones labeled APALATHEAN MOUNTAINS. Frolic with those naked Indians pictured in that sparkling stream. What did I have to lose? This whole jaunt down I-85 was a giant crapshoot anyway. Might as well hitch a ride with the past, if only to see where it took me. And at that moment it was taking me to an Indian village illustrated on that map with thatched huts. These days, that same spot is known as Hillsborough, North Carolina, and if I could have stuck to the precise route of the trading path, it would have led me to within a block of Gilbert Bland's jail cell.

One of the first Europeans to visit the place Bland now calls home was John Lederer, a German-born adventurer whose geographical observations were later incorporated into the Ogilby-Moxon map. In June 1670, Lederer arrived at what was then a Native American settlement along the Eno River. He found the people of these parts to be "covetous and thievish, industrious to earn a peny [*sic*]"—which also struck me as a pretty fair description of the town's most recent resident. But the sad truth was, I didn't really understand the map thief a whole hell of a lot better than Lederer understood those Indians. Even after all these months, I still knew only the *what* of his story, not the *why*. So that's what I had come to find out. I had this idea that, away from the influence of his lawyers, Bland might be willing to talk with me, especially if I showed up unannounced at his appointed house of correction. Anyway, it was worth a try.

He was there to face state charges stemming from his heist at the University of North Carolina in nearby Chapel Hill. In addition, he still had to contend with state charges in Delaware, as well as the federal indictment in Virginia. At the time—July 1996—none of these cases had been resolved, so Bland, unable to make his $75,000 bond, was biding his time at the Orange County Jail in downtown Hillsborough. I went straight there, edgy to make contact with him, but was promptly told that there were no visiting hours on that day. To see the prisoner I would have to return the following morning. In the meantime, a couple of bored-looking jailers agreed to let him know that I was in town.

"Historic Hillsborough," as the travel brochures call this village of five thousand, is packed with late-eighteenth- and early-nineteenth-century buildings, making it something of a living museum, half Williamsburg, half Mayberry R.F.D. Wandering around town, the crape myrtle in full bloom, I had a renewed sense of being enveloped into the past. On one side street was a still-functioning inn, built in 1759 and rumored to have been frequented by such notables as the British general Charles

Cornwallis, who led his troops over the trading path during the waning months of the American Revolution. I drifted inside and a few moments later found myself settling into a room just a block down from Bland's jail cell.

That night it happened again. Lying there in the dark, the window air conditioner murmuring *aum* like some Hindu mystic, I stumbled into yet another locale from the Island of Lost Maps. This time I wasn't scampering down a footworn Piedmont trail but rushing headlong into the South American jungle, toward a place labeled "Manoa & el Dorado." Blame this weird waking dream on Sir Walter Raleigh, who had been at the back of my mind all day, his memory being nearly unavoidable in a state where even the capital city (to which I had driven for dinner a couple of hours back) bears his name. But if Carolinians remember Sir Walter for organizing the first English settlements in North America, the ill-fated Roanoke colonies of the 1580s, I was more interested in a different venture, his 1595 search for a legendary land of gold. He did not find it, of course—but that did not hinder him from writing a fanciful account of his trek, in which he claimed to have come within a short distance of "the great and Golden Citie of Manoa (which the Spanyards [*sic*] call El Dorado)." He was also the first explorer known to have put Manoa on a real map—and, thanks to Raleigh's fame and standing, his views commanded the respect of other cartographers. Jodocus Hondius, apparently using Raleigh's 1596 map as a model, included Manoa in his 1598 map of South America. In 1599 the Dutch publisher Theodore de Bry issued a Latin translation of Raleigh's book and included a map showing the city of gold with the caption "Manoa or Dorado, regarded as the greatest city in the entire world." The mythical metropolis continued to appear on many seventeenth-century maps—including Gerard a Schagen's *Totius Americae Descriptio,* into which I was at that moment fading—and would remain on some works until as late as 1808.

In addition to inventing imaginary places and people (including a tribe of headless men described in his El Dorado narrative), Raleigh had a habit of reinventing himself. He was, as the writer Robert Silverberg put it, "a man of many characters"—one of several traits he had in common with Bland. Each man was in his forties when he embarked on his quest for sudden riches. Each appeared to be on the verge of a financial collapse. ("There are many possible readings of his El Dorado quest," wrote Charles Nicholl in *The Creature in the Map: A Journey to El Dorado,* "but let us not forget the simple ones: for instance, that [Raleigh] was, or was going, broke.") Like Bland, Raleigh had a penchant for shady scams (having dabbled in both extortion and piracy). Like Bland, he had been to prison before and would go there again.

Yes, but I knew that I, too, shared something with old Sir Walter. For all his bluster, Raleigh seems to have sincerely believed his expedition ended just shy of a place where "the graves have not beene opened for gold, the mines not broken with sledges, nor [the] Images puld down out of [the] temples." The hope of reaching that promised land never left him. Haunted by an unfinished quest of my own, I could relate to that. I could put myself inside his head, dream his dreams, imagine how it feels to emerge from the dark jungle into a lush valley, "the deare crossing in every path, the birds towardes the evening singing on every tree with a thousand several tunes, cranes & herons of white, crimson, and carnation pearching on the rivers side, the ayre fresh with gentle easterlie wind, and every

THEODORE DE BRY'S SIXTEENTH-CENTURY MAP OF GUIANA SHOWS "MANOA OR DORADO . . . THE GREATEST CITY IN THE ENTIRE WORLD" (SEE INSET), TO SAY NOTHING OF A HEADLESS MAN AND AN AMAZON.

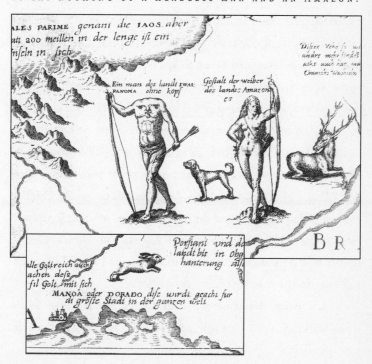

stone . . . eyther golde or silver by his complexion." And as that map rushed in at me, filled the corners of the room like darkness itself, I could almost see the glow of a "great and golden Citie" somewhere off in the misty hills. So there I was, chasing El Dorado all the way into the deepest part of sleep, and then it was morning, and there I was, off to visit Gilbert Bland.

The bored-looking jailers got straight to the point. He would not see me. This was not a surprise, and somehow not even much of a disappointment. If anything, my immediate reaction was a mild sense of liberation. I could finally relax: the quest was finished. Or so I now imagined. Later, I would begin to wonder if those bored-looking jailers had even bothered to pass along my message. Later, I would start to believe that, even if they did, only Bland's legal circumstances held him back. Later, I would begin to hope that he'd agree to speak with me once his court cases were resolved. But all that anxiety was for the future; for the moment, everything seemed simple. I had but one thing left to do in town. A tourist brochure reported that archaeologists digging along the Eno River near downtown Hillsborough had discovered Native American settlements going back to the Stone Age. This, I realized, was that village I had seen on the Ogilby-Moxon map—the original destination point of my pilgrimage into the past. If I accomplished nothing else on this sorry expedition, at least I would complete that journey. I would give myself over to the map one more time.

The dig site, it turned out, was only a few hundred yards from Bland's jail cell, sitting in a fallow field at a bend in the Eno

River, right down from a row of suburban-style homes. I walked out toward it, passing a rusty International Harvester tractor (itself so antique it could almost pass for the unearthed artifact of some forgotten culture), then pushing through a preternaturally large spiderweb, which hovered across my path like a phantom. The whole place, in fact, had a ghostly aura, an air of being more absent than present. The archaeologists had apparently gone home for the summer, covering their work with neat mounds of dirt that reminded me of fresh graves— the only outward signs of a Native American presence that dates back more than ten thousand years. So what happened to this place? Why its sudden demise in the early 1700s? The list of answers, I would later learn, included many of the usual suspects, such as warfare with other Indians and the devastating spread of European diseases (which, according to one early eighteenth-century chronicler, reduced some Piedmont-country populations by more than 80 percent in fifty years). But, standing there on that empty plain, I was already beginning to understand the role of another factor, less obvious but perhaps no less important: a clash of cartographies. When the adventurer John Lederer had rambled down the Great Indian Trading Path in 1670, for instance, he was not just sight-seeing. Working under the auspices of Virginia's governor, Lederer was at the vanguard of a systematic effort to appropriate land—an effort in which maps often played as big a role as guns. As the scholar Mark Monmonier wrote in *Drawing the Line: Tales of Maps and Cartocontroversy:*

Indigenous Americans communicated information about space and places through folktales, gestures, dances, and ephemeral drawings, but theirs was not the cartography of commerce, navigation, and warfare. Land ownership in the profane European sense of buying, selling, inheriting, recording, and taxing was an alien concept. American Indians, who considered land sacred and not "ownable," never developed a formal cartography focused on boundaries and surveys. This lack of maps—really a lack of what the European invaders recognized as maps—was one of many technological disadvantages that made the conquest of the New World not only quick and easy but also morally right in the minds of the colonists and their priests.

Even the map that had brought me here was made with pilferage in mind. Ogilby and Moxon's work was used as an advertisement to attract English settlers to the new province, whose colonization had begun only a few years earlier when King Charles II granted Carolina to eight wealthy lords proprietors. It was no accident that the new owners' names appeared all over the map (Ashley River, Cape Carteret, Clarendon County, Albemarle River, et cetera), obliterating the Native American identity of those places in the same way that Russian Communists would later try to purge the memory of their imperial predecessors by renaming perhaps half of the seven hundred thousand towns and cities in what became the Soviet Union. With the power to manipulate the land from afar—to hype it up, slice it up, divvy it up—the Ogilby-Moxon map gave colonizers a tool of theft every bit as sharp-edged and efficient as Gilbert Bland's razor blade. Indeed, it now occurred to me that not only did this one small document encapsulate the whole long saga of New World exploitation, but that when Bland laid his hands on it, he was reenacting that story in a very poignant way. His victims had almost the same relationship to their maps as the people of this site had to their land. For librarian and Indian alike, these were "sacred and not ownable" resources—a lovely ideal, which, like most lovely ideals, proved easy to exploit. I came here to find out why Bland did what he did. Well, maybe it was as simple as this: because he could. Until that moment I had viewed him as something of a cultural freak, a man whose actions were outside the normal course of human events. But now, as I stared across that haunted little patch of earth, he seemed very much a part of the landscape.

I HAD PACKED MY BAGS, DROPPED OFF MY KEYS. I HAD finished my breakfast, hopped into the car, and, just for the hell of it, cruised past the jail one last time. There was a plane to catch the next morning, hundreds of miles away. It was time to crank the AC and hit the road. But wait. Did I just see what I thought I saw? There, on Route 109 of my official North Carolina State Transportation Map, south of Healing Springs, north of Mount Gilead, east of Misenheimer, west of Spies, smack-dab in the middle of the Uwharrie National Forest, was a spot called Eldorado. Eldorado, the place that is always farther on. The land that exists only in a person's mind. The destination to which, as Charles Nicholl observed, "there is only the journey, the approach toward something that you cannot reach, something . . . that you dare not reach." Or . . . well, maybe I dared, after all. True, it was at least a hundred miles in the wrong direction. And true, I couldn't come up with one rational reason to undertake yet another wild-goose chase. But like some born-again who gets rich by heeding advice from random Bible passages, I was beginning to believe in the power of cartomancy. Maps had been talking to me even more than usual. Let this one be your Ouija board, I told myself. Close your eyes and ask whether to go or stay. Open them again. What do you see? A place not far from Eldorado: the town of Whynot.

And so I was off, like so many dreamers, desperadoes, and idiots before me: Raleigh, whose search for the golden city cost him his son and then his own head; Gonzalo Pizarro, whose men were reduced to eating horses, dogs, and saddle leather before abandoning their journey; Philipp von Hutten, whose expedition became what one chronicler called a "theater of miseries," the final act of which was the leader's decapitation with a blunt machete; Gonzalo Jiménez de Quesada, whose misadventure left 250 Spaniards and nearly 1,500 Indian porters dead and "achieved nothing," according to the survivors; and, finally, the infamous Lope de Aguirre, the crazed leader of a wilderness mutiny who declared himself "Wrath of God, Prince of Freedom, King of Tierra Firme," before his own followers killed and quartered him, then put his corpse on display in sections. "The reports are false," Aguirre wrote about El Dorado, not long before his death. "There is nothing . . . but despair." Not exactly a good omen for my little road trip—but if anything can be said for certain about the golden city, it's that the place brings out the optimist in a person. And so I turned up that country-and-western station and hit the pavement, just like Edgar Allan Poe's "gallant knight," who

> . . . *journeyed long,*
> *Singing a song,*
> *In search of Eldorado*

I followed the Great Indian Trading Path, aka I-85, west to Burlington, then cut down Route 62, past Julian and Climax, to Route 220 south. Everything was going just fine—but then a kamikaze rainstorm swooped out of the bright sky, making it difficult to keep track of the car in front of me and slowing my pace to a crawl. By the time I stopped for a dried-out chicken sandwich at the Hardee's in Randleman, I was beginning to lose steam.

> *But he grew old—*
> *This knight so bold—*
> *And o'er his heart a shadow*
> *Fell as he found*
> *No spot of ground*
> *That looked like Eldorado.*

I pushed on, past Asheboro, to Route 134 south, then followed that two-laner to Troy, a bleak town of textile and log mills. When I finally got on Route 109 and entered the Uwharrie National Forest, the afternoon and my patience were both seriously on the wane.

> *And, as his strength*
> *Failed him at length,*
> *He met a pilgrim shadow—*
> *"Shadow," said he,*
> *"Where can it be—*
> *This land of Eldorado?"*

> *"Over the Mountains*
> *Of the Moon,*
> *Down in the Valley of the Shadow,*
> *Ride, boldly ride,"*
> *The shade replied—*
> *"If you seek for Eldorado!"*

Oh, I was riding boldly, all right—thanks to another caffeine infusion from the kindly Dr Pepper. Somehow, though, I had managed to bypass the Mountains of the Moon and the Valley of the Shadow, winding up instead at a general store in the tiny burg of Uwharrie. Where can it be—this land of Eldorado? Well, shit, son. It's just up the road.

The first thing I learned about the place was this: even when you find it, it's not really there. The Macedonia Methodist

Church, the Adams Egg Farm, a few forlorn houses and ramshackle trailers—welcome to Eldorado (pronounced El-duh-RAY-doh by the locals), where the streets are not paved with gold and some are not paved at all. The most impressive edifice by far was the Eldorado Outpost, a new convenience store with deer and fish trophies on the walls, groceries and "Quality Gun Care Products" on the shelves, and Seen Better Days pizza under the heat lamps. I asked one of the girls at the cash register about the history of the town.

"Um, it's not really a *town*," she said with a giggle, before searching behind the counter for a photocopied "minihistory" by the county historical society. It informed me that Eldorado had been the site of a gold discovery in 1885. Whatever small riches were made here obviously did not last. Eldorado, I realized, was nothing but the last remnants of somebody's doomed schemes and blind greed. And, suddenly, I got a funny feeling that I'd been there before.

GILBERT BLAND OUTSIDE THE
COURTHOUSE IN HILLSBOROUGH,
NORTH CAROLINA, ON JULY 1, 1996.

CHAPTER 13

Mr. Bland, I Presume

EVEN AFTER ALL THE SETBACKS, I KNEW OF one last spot where I was sure to come across Bland. As his case shuffled toward completion in three jurisdictions, the map thief's calendar had begun to fill with events he could hardly refuse to attend. And so, on a bright, brisk day in December 1996, I found myself at a federal courtroom in Charlottesville, Virginia, where he was due to appear at a sentencing hearing. I would finally lay eyes on the man.

I had been awaiting this event with a kind of anticipation that bordered on frenzy. At times, I felt nearly as anxious as the explorer Henry Morton Stanley, who, on September 19, 1871, scribbled these words from somewhere deep in central Africa:

> I have taken a solemn, enduring oath, an oath to be kept
> while the least hope of life remains in me, not to be

tempted to break the resolution I have formed, never to give up the search, until I find Livingstone. . . . And something tells me, I do not know what it is—perhaps it is the ever-living hopefulness of my own nature, perhaps it is the natural presumption born out of an abundant and glowing vitality, or the outcome of an overweening confidence in one's self—anyhow and everyhow, something tells me . . . I shall find him, and—write it larger—FIND HIM! FIND HIM! Even the words are inspiring. I feel more happy. Have I uttered a prayer? I shall sleep calmly to-night.

Stanley's agitation was understandable. Having marched more than five hundred miles through vast swamps and dense jungles in search of the lost geographer-missionary David Livingstone, enduring 128-degree temperatures, torrential rains, unfriendly locals, mutinous underlings, bloodsucking tsetse flies, malaria-carrying mosquitoes, and beetles as large as mice, he was, at the time of his writing, half-loopy from "fevers without number." So what was my excuse? It was clear by now that my search had led me into dangerous emotional territory. And so, having invested many months in trying to figure out what made the map thief tick, I now began also to look at my *own* motivations. If I was going to keep pursuing this frustrating quest, I thought I owed it to myself—and perhaps to Bland as well—to figure out exactly what I was doing.

I realized that in spirit, at least, my search resembled a kind that popped up fairly often in the annals of discovery. As far back as Homer's time, when Telemachus set out after his long-lost father, Odysseus, much of our world has been mapped by explorers who were not actually seeking new lands but hunting for elusive human beings. In the Middle Ages, for instance, generations of European voyagers headed to Asia and later Africa in search of Prester John, an ever-elusive Christian king whose vast and wealthy dominions were said to have "no theft nor sycophancy nor greed nor divisions." We may now snicker at maps such as *The Kingdom of Prester John,* published by Ortelius in 1573, depicting a huge African empire. Nonetheless, the hunt for this mythical potentate unquestionably helped to end medieval Europe's isolation and spark the Age of Discovery.

Nor were other eras immune from what Eric Leed, author of *Shores of Discovery: How Expeditionaries Have Constructed the World,* described as "the Lost-Boy Complex." In 1805 the Scottish explorer Mungo Park vanished along the Niger River. He was never found, but those who attempted to track him down dispelled many geographical myths about the West African interior. Likewise, after British rear admiral Sir John Franklin disappeared during his 1845 attempt to find the elusive Northwest Passage, more than twenty expeditions were sent out in search of the mariner and his men (all of whom, it turned out, had perished). "The Franklin affair was the greatest disaster in the history of Polar exploration," wrote Peter Whitfield, "yet in drawing dozens of search-parties to the region in the 1850s, it was responsible for completing the map of the Arctic."

My own quest had many motivations, of course, not the least of them being money and professional pride. Like Stanley, who was sent to Africa as a reporter for the *New York Herald*, I was first and foremost a journalist on assignment. There was also an intellectual challenge. The more Bland had avoided me, the more I had learned to rely on my own resourcefulness and research skills to patch together the details of his life. But something else kept me going, an almost instinctual drive, seemingly linked to the allure of the unknown. It was beginning to remind me an awful lot of this Lost-Boy Complex. And as I read through the accounts of explorers, trying to make sense of the phenomenon, a curious pattern began to emerge. In a striking number of cases, the searcher—consciously or not—began to take on traits of the missing person, finishing his quest, meeting his same fate, or otherwise merging with his being. Stanley, for instance, had suffered a lifelong crisis of identity. An illegitimate child who was abandoned by his mother and never knew his father, he changed his name from John Rowlands to Henry Stanley, after a man whom he idolized as a father figure. His hunt for Livingstone was likewise a search for self, as if he were driven not only to *locate* the great adventurer but also to *become* him in some way. Stanley himself hinted at this very notion in his journal. In the older and more famous man's presence, he wrote, "I . . . begin to think myself somebody, though I never suspected it before."

Livingstone was, in many ways, worthy of emulation. As a physician he dedicated himself to using modern medicine to alleviate the suffering of the African people; as a Christian missionary he battled the slave trade; as an explorer he made sweeping contributions to geographical knowledge of the continent. I could see why Stanley would be drawn to him. But just what was I trying to find in a man like Gilbert Bland?

*F*ACES HAVE OFTEN BEEN COMPARED TO MAPS. Ptolemy himself described mapmaking in terms of painting a portrait: in both disciplines, he wrote, one must be concerned not only with the whole head (the entire world) but with individual features (particular places). Likewise, the Dutch master Jan Vermeer—a contemporary of Golden Age cartographers such as Joan Blaeu, Jan Jansson, and Frederick de Wit—repeatedly counterposed the human face with the face of the Earth. In works like *The Art of Painting, The Soldier and a Laughing Girl, Young Woman in Blue, Young Woman with a Water Jug,* and *The Geographer,* Vermeer placed the face of his subject in front of a map—as if to show that the former charted the inner world just as the latter depicted the world at large. And this same metaphor was employed by any number of writers, including William Shakespeare in *Henry VI, Part II:*

> *In thy face I see*
> *The map of honor, truth, and loyalty*

So what sort of map, I wondered, as I sat anxiously in that courtroom, would I see on the face of the defendant? By now several photos of Bland were etched into my memory: the waif-like face from his first mug shot in 1968, cheeks angelic but eyes empty; the noticeably hardened face from a post-Vietnam mug shot in 1971; the dazed face from another arrest in 1972, hair shoulder-length and shaggy, mustache full and walrusy, à la David Crosby; and, finally, the face of a middle-aged criminal, mustache graying, hairline receding, eyes peering past the camera defiantly, one brow arched at a foreboding angle. But while those photos charted the changes that had taken place in Bland's outward appearance, they told me little about his soul: poker faces, every last one of them. If nothing else, I hoped that this long-awaited chance to observe him in person would allow me to decipher his features, as Stanley had done with Livingstone:

> I found myself gazing at him. . . . Every hair of his head and beard, every wrinkle of his face, the wanness of his features, and the slightly wearied look he wore, were all imparting intelligence to me—the knowledge I craved for so much ever since I heard the words [from my publisher], "Take what you want, but find Livingstone."

As the appointed hour approached, the courtroom filled with lawyers and clerks. There were no well-wishers or family members seated in the gallery—just a few reporters quietly readying their notebooks. And then, there he was, shuffling to his seat in blue prison scrubs. In a similar moment of encounter, H. M. Stanley had professed a desire to "vent [his] joy in some mad freak, such as . . . turning a somersault." But, despite months of anticipation, I now felt nothing but a vague sadness, whether at Bland's meager appearance or my own grandiose expectations, I could not be sure. Although I had hardly imagined a robust individual, I was taken aback by how sunken he looked. Five feet nine inches tall, with reddish brown hair that was beginning to go gray, he seemed hunched and frail, as if his body belonged to an old man. Yet he also had the look of someone far younger than his forty-seven years: without the mustache I had seen in so many of his photos, he seemed almost childlike. His skin was pulpy and lusterless, as though never exposed to sunlight. And his expression? Empty. Blank. Bland. It was simply not a face, I now realized, that gave itself away. The only thing vibrant about it was his piercing hazel eyes. A couple of times he leaned back and sneaked glances at me as I stared at him. I'm not sure he even knew who I was, but his look was not friendly. I could imagine how Jennifer Bryan must have felt when he darted his "surreptitious" eyes at her on that fateful day in the Peabody Library. It was the gaze of a man who intensely disliked to be observed.

Prison had not been good to the map thief. Earlier in his incarceration, while staying at the Albemarle-Charlottesville Regional Jail, he'd written the U.S. district judge to complain about having to live in crowded conditions with a number of violent criminals. He attempted to pass the days by reading, he said, but had trouble concentrating in a place where the television constantly blared, "with RAP music videos, cartoons, and wrestling. . . .

"I have tried, with the help of anti-depressant medication . . . to maintain and cope," he wrote, but "the stress is unbearable." Noting that two other inmates had hanged themselves since his arrival several months earlier, he said he was worried about "retain[ing] my sanity."

Knowing that Bland had written similar letters from prison in the past, I could not help but feel a certain skepticism about his claims of mental illness. But when the hearing got under way in court that morning, his attorney argued that the map thief's troubled emotions were indeed at the heart of this case. In urging a light sentence for his client, the Roanoke-based lawyer Paul R. Thomson, Jr., said Bland's "pattern of problems" was "largely triggered by depression, a very common problem with post traumatic stress syndrome and something that we saw a whole lot of during the Vietnam era." Thomson also cited Bland's financial troubles as motivation for the crimes. "He did it in large measure because his business that he owned in Maryland . . . was failing," he said, adding that Bland "could not feed his family." Thomson assured the court that his client would remain in an outpatient treatment program once he returned home to Florida. "He recognizes that this was a singularly poor judgment. . . . This has been an emotionally traumatic incident for him, for his family."

Any hopes I had entertained of Bland himself offering insight into the crimes proved sadly optimistic. Speaking in a meek voice that occasionally snagged with emotion, he gave what amounted to a stock repentant felon speech: "The first thing I'd like to say, Your Honor, is that I'm truly sorry for what

I've done. I'm ashamed of myself. I'd also like to say, Your Honor, that in the year just gone by, I've had a lot of time to think about why this has happened and also . . . about my responsibility to my family, to my wife and children when I get back home. I have a lot to make up to them. I'm forty-seven years old now and I realize that I can't make mistakes like this anymore. . . . It will never happen again."

Then he quietly slouched back to the defense table and seemed to melt into his chair, the soul of inconspicuousness.

UNITED STATES DISTRICT COURT JUDGE JAMES H. Michael, Jr., a tall, white-haired man with hawkish eyes, a resonant voice, and the stately air of an old-fashioned Southern gentleman, wanted Bland to know how much the thefts had offended him. "In the court's view, the loss is irreplaceable and unmeasurable," he told the map thief at that hearing. Nonetheless, a plea bargain had been signed, and the judge said he made it a rule to honor plea bargains. Accordingly, he sentenced the defendant to eight months in prison—time already served—and ordered him to pay seventy thousand dollars in restitution to the University of Virginia and Duke University. (Bland would later contest that amount, noting that the damages he had caused were easily remediable, since many of the maps could simply be glued back into their atlases—in response to which Judge Michael would accuse him of "a persistent misunderstanding of the gravity of the offense.")

Ten days after his sentencing in Virginia, Gilbert Bland was in court once again. Like their federal counterparts, North Carolina prosecutors had agreed to a plea bargain that would have let the map thief off with time served. This time, however, the judge had other ideas. Noting that Bland had stolen twenty-six maps and documents worth an estimated twenty thousand dollars from UNC–Chapel Hill, Superior Court Judge Robert Hobgood rejected the deal, insisting that "the penalty is not severe enough for what this man has done." It was a gutsy stand, so much appreciated by librarians that it earned Hobgood an award from the Smithsonian Institution's 1998 National Conference on Cultural Property Protection. But the practical impact of the decision proved minimal: even with the judge throwing the book at him, Bland would end up serving only five additional months in prison. After receiving a two-year suspended sentence in Delaware, where he was ordered to pay nine thousand dollars in restitution for the maps he took from the university, Bland was ready for release, having spent a total of less than seventeen months behind bars.

As the news of Bland's imminent freedom spread via the Internet, librarians from around the globe seethed at what they viewed as a laughably light sentence. "In my opinion, he should have got twenty years hard labor. . . . Those who talk their way into our confidence and then betray our trust are lower than the dirt," snarled P. D. Hingley, librarian for the Royal Astronomical Society in London.

"Since civility prevents us from chopping off his hand, any such thief should be stuck away for a year for each dollar's worth of stuff he stole and for each dollar it cost to apprehend, prosecute, and incarcerate him," added Sidney E. Berger, head of special collections at the University of California, Riverside.

Yet, despite such get-tough rhetoric, the librarians knew that their own peers were partly to blame for the situation. Of the nineteen institutions allegedly hit by Bland, only four had pressed charges. The others had simply let the matter drop. Why the unwillingness to act? "I've asked that same question many times, and I haven't gotten much of an answer," said Cynthia Requardt of Johns Hopkins, where Bland was apprehended but never charged. "I think it fell through the cracks. The legal counsel didn't think that there would be much gained by it, everyone else was waiting for the legal counsel to take the first step. . . . It dragged on so long, and then Virginia prosecuted, and then Delaware and North Carolina, and so the feeling became, Well, it's been taken care of."

True, Bland's federal plea bargain would have made additional state cases against him difficult to pursue—but by no means impossible. "Other libraries would have been able to prosecute if they had really pushed," insisted the FBI's Gray Hill.

Moreover, the U.S. plea bargain did not shield Bland in Canada, where he had allegedly walked off with a total of almost forty maps from the University of British Columbia in Vancouver and the British Columbia Archives in Victoria. Nonetheless, no charges from north of the border ever materialized. "I talked to the Royal Canadian Mounted Police about what our chances would be," said Brenda Peterson of the University of British Columbia. "And they said that in order to bring any sort of charges against him, we'd get into international law, and that would be very complicated. They didn't hold out much hope that the amount of effort and money that would be required to actually charge him in Canada would be worth it. Let's just put it this way: they did not urge me to pursue it."

And so, on May 23, 1997, less than a year and a half after entering jail, Bland strolled out into the spring air, a free man. In a curious twist of fate, his last days behind bars had been spent at Fort Dix, New Jersey—a federal prison established in 1993 on the site of a former military training facility that Bland knew all too well. It was into Fort Dix he had walked for induction into the Army in 1968, and it was out of Fort Dix he had come upon leaving the service in 1971, anxious to put the past behind him. Now he had reason to hope for another new start. The police and FBI were no longer after him. His victims had moved on to other concerns. And, as the weeks wore on, most of the media forgot about him, too. I was the last one on his trail.

*H*ELLO."

"Hi, Mr. Bland."

"Yes, who is speaking?"

"This is Miles Harvey calling."

I had waited more than eight months after his release to telephone him. His lawyers had told me he was not pleased with the publication of my magazine article and had given me no reason to believe he was any happier about the possibility of a book. Still, I hoped that, with the passage of time, I might find him in a reflective mood. I was mistaken.

"First of all," he said curtly, "do you mind if I record this conversation?"

"No. Do you mind if I go ahead and record it, too?"

"Be my guest."

"All right, thanks."

I was surprised at how shaky my voice sounded—and how calm his seemed. It was also deeper than I remembered it, and his slight New Jersey accent seemed to give it a toughness. There was a hint of menace in it, too. Even from the first few words I sensed that it was a voice that could persuade, a voice that could intimidate and injure—a voice, in short, entirely unlike the broken and barely audible one I had heard in court.

He demanded to know how I had found his phone number. I told him the source—a document on the public record—then fumbled for words. "I just wanted to introduce myself to you."

"Let me tell you something, first. I do not wish for you to communicate with me in any way and in any manner," he said flatly, as if reading from a prepared statement. "There are laws in the state of Florida that protect people. I do not wish for you to communicate with me any further. If you do, I will try to bring criminal charges against you, and I will certainly bring civil charges against you. Do you understand that? Is that clear?"

"I understand what you just said, yeah."

"Okay."

Although his legal threats were unfounded if not insincere, it was obvious the conversation was not going well. I knew this would be my last chance.

"Now, what I'd like to request is that we at least meet in person," I said. "I think you should meet me, even if you don't want to talk to me. I—I'm a good guy. I'm interested in getting to know whatever part of this story you want me to get to know. I just want to give you every opportunity; I just want to be fair here. It's part of my job. I have absolutely nothing against you. I have absolutely no interest in pestering you or invading your privacy. I'm just trying to do a good job as a reporter. Can you understand that?"

"Uh, you've been warned, and I suggest that you check Florida statutes regarding stalking before you try to communicate with me again. As I said before, I'm warning you and I have a tape recording of this conversation. If you continue to harass me or attempt to communicate with me in any way, I will file charges with the police and I will charge you in a civil action. Do you understand that?"

"Yep, I understand what you said."

"Is that clear to you?"

"It's absolutely clear. I'm still hoping to talk you out of it . . ."

"I have nothing more to say to you."

Then came a silence. A familiar silence. A final silence.

Seventy years ago, when her husband was con-victed of embezzlement and shipped off to federal prison in Leavenworth, Kansas, my grandmother faced an agonizing de-cision: how to break the news to her young daughter. Family lore has it that my grandfather was guiltless, save for refusing to snitch on a friend who worked above him at the bank (and who later shot himself to death at a church in our hometown of Downers Grove, Illinois). But my grandmother knew that, even if her husband was innocent, his departure would devastate their youngest child, then about six years old. A levelheaded and resourceful woman who taught many generations of ele-mentary school students, my grandmother was usually the type of person who knew precisely what to do in a crisis. This time, however, she felt overwhelmed, paralyzed by a combina-tion of grief, loss, and the prospect of sudden poverty at the height of the Great Depression. For her teenage stepson, from whom news of the trial could not be kept, the situation was bad enough: he would have to abandon his college plans and look for work in the government-run Civilian Conservation Corps. But how could she possibly hope to explain the tragedy to her daughter, a sensitive little girl who idolized her father?

Desperate for help, my grandmother turned to an associate at her school and another friend with some supposed expertise in child psychology. Their advice was simple, rational, and, it turned out, tragically wrongheaded. Don't tell the girl anything, they said. There is no need to mention the word *prison;* a young child could not cope with the shame. Let her believe that her father has just gone away on a trip. Tell her he'll be home any day.

And so for years my mother went to sleep every night wondering whether the next morning would bring her father back from his journey into the shadows. She does not remember how long he was gone. She does not remember asking anyone where he went. She does not remember trying to picture him in a faraway place. He had simply disappeared, faded into some misty terra incognita, a place all the more terrifying because it was entirely undefined and undiscussed, as distant and incomprehensible as death itself.

He finally did return one day—but, in many ways, she waits for him still. I grew up with a mother who lived in constant fear that her loved ones would simply vanish. If her husband tarried too long on a trout stream, as he was prone to do, or if her teenage sons remained too long at a party, as was their wont, this normally vibrant and easygoing woman would dissolve into a panic. On nights when I came home far past my promised hour of return—there were too many of these, I regret to say—I would often find her sitting by the phone, sobbing, her terror so manifest it seemed to shrivel her six-foot frame. Yet if my grandfather's disappearance left her with a crippling fear of the unknown, it also left her with a very different, very healthy kind of dread: a fear of *not knowing.* My mother's fierce curiosity—a lifelong quest for answers, to both the world's problems and her own—has been, I think, her salvation, a sweet revenge on the forces of uncertainty that haunted her childhood.

I inherited, in equal and all-too-generous portions, her inquisitiveness and her skittishness. And, in retrospect, I realize how much these two contradictory traits colored my investigation of the map thief—one of them driving my search, the other filling me with trepidation about what I might find. Earlier in this book I wrote that Bland seemed such a stranger as to be "beyond my boundaries." I now understand that his remoteness, both psychological and physical, was the main part of my interest in him. I cannot say whether in seeking him out I was somehow looking for the grandfather I never met, a man who, like the map thief, had fallen from social grace in middle life. All I know is that, with each passing month, it seemed that I was searching less for an actual person named Gilbert Bland than for some dark and unexplored part of my own existence. And then one day, to my fascination and horror, I caught a glimpse of it. Him. Me. The thief. My years of attempting to get inside Bland's head had been a failure, I realized. I had not penetrated his thoughts, only imitated his actions—sneaking around the edges of his life just as he had crept around libraries, slicing away little pages of his past, then secreting them home. Although Bland's threat to sue me for stalking was all bluster, I could not really blame him for feeling misused. His sense of violation, in fact, must have been very much like that of his own victims.

And while I agreed with those victims that the judicial system had let Bland off easy, I also understood that punishment is not always measured in jail time alone. From my own family's experience, I was tempted to believe Bland's lawyer when he stood in federal court and told Judge Michael: "I think the biggest punishment that he will face is having to go back and face his teenaged son and explain to him how he made this decision."

Moreover, I knew there was yet another punishment awaiting Bland—one that I myself would be bound to administer.

I had hoped my search for the map thief would end with a discovery, some thunderbolt of truth that threw light upon all his actions. That never happened. Yet I came to understand that, at its most fundamental level, a discovery has less to do with revelation than with declaration. Just as the word *explore* comes from the Latin for "to cry out," discovery is the act of *making known*. Christopher Columbus was not the first to arrive in America: his genius was in introducing the New World to the old one. "To discover is to draw the veil," observed the Columbus scholar Mauricio Obregón. "It is not to run into something and keep it to oneself. It is to push back the frontier of infinity and to pass the news on to posterity."

But every discovery is also an act of appropriation. After all my years of research, the only thing I could say about Gilbert Bland with absolute certainty was that he cherished his anonymity. I had seen him nourish it through a succession of aliases, deploy it ingeniously in one rare books room after another, guard it like a pit bull against all outside forces. Even the courts had not taken it away from him. But I would. This book would. It was not that I meant him any harm: if anything, my feelings of compassion toward him had only grown over the years. But I had come to realize that, no matter what my intentions, the telling of his story would doom Bland to what was, for him, a most terrible fate: to be known.

I was confident that the penalty fit the crime. If nothing else, calling attention to Bland seemed likely to deter him (and other would-be library crooks, I hoped) from future pilferage. Nonetheless, over time I grew increasingly ambivalent about my pursuit of the map thief. I never considered abandoning the project entirely, but I did feel a growing moral fatigue, a creeping doubt about why I continued to rifle through Bland's life for ever more obscure details—especially after it was clear that, without his cooperation, those details would never add up to a complete picture of the man.

One last truth about the act of discovery: it's easy to get lost in the wilderness. Henry Morton Stanley went into the African jungle driven by curiosity, a reporter in search of a scoop. He came out years later, a cynical adventurer for hire, who, as a mercenary of Belgium's brutal King Leopold II, established the Congo Free State, a vast colony of forced labor, torture, and terror where literally millions of people perished. Perhaps it was also curiosity—"a genuine interest in and affinity for" old maps, in the words of one defense attorney—that first lured Gilbert Bland into the peculiar terrain of the rare books room. But curiosity, the desire to experience and understand a thing, has an unfortunate way of devolving into a need to possess it, to conquer it, to make it submit to one's will. And when that happens the explorer becomes the exploiter. I had tried to prevent that particular tragedy from befalling my own adventure. But one day, while transcribing the tape of the phone conversation with the map thief, I heard my voice strain at the words "I'm a good guy." It struck me that I had been trying not to convince Bland but to reassure myself. And I knew that I, too, was beginning to lose my way.

We rarely reach our destinations, at least not the ones we set out to find. More often, we arrive one day at a place unfamiliar and unexpected, where all roads suddenly seem to converge and something—the ineffable smell of fate or the stench of defeat or (likeliest of all) sheer exhaustion—tells us the journey is over. In my case it was the knowledge that I had wandered too far. We do not take trips so much as they take us, and this one had started to transform me into someone I did not want to become. It was time to head home.

FOR WRITERS A PLOT IS A SERIES OF EVENTS; FOR cartographers a plot is a map. By either definition, this plot is now at an end. That does not mean it is complete: no plot ever really is. But there comes a point at which you must let the ink dry and step away from the drafting table, hang your work on the wall and see what's there.

Any attempt, whether in writing or mapping, to fit our three-dimensional world onto the flat page leads to distortions. Some elements receive too much space, others too little; still others become warped almost beyond recognition. As I examine my plot I see not only that it suffers from problems of projection but that much of its geography is speculative. Mountains undoubtedly have been sketched where molehills exist, and vice versa. Such guesswork is inescapable when one must depend almost entirely on other people's observations, often nothing more than a shoreward gaze from the crow's nest of a ship. And then, of course, there is the subjectivity of the plot maker himself. Although I have tried to be exacting, all cartographers occasionally succumb to myth and prejudice, thereby transforming entire landforms into Rorschach inkblots that reveal more about the draftsman than the terrain.

"Nature conceives of innumerable things, of which those known to us are fewer than those not known," wrote the fifteenth-century mapmaker Fra Mauro, "and this is so because nature exceeds understanding." To which I would add: especially human nature. As I study my plot, I see a land whose coastline is rendered with some degree of accuracy but whose interior, despite my best efforts, is still stamped *Lands Unknown*. I had hoped to do so much more. I had hoped to put things in exact relation and scale, to get at a single reality. Instead, I see only rumination and whimsy, arabesques of uncertainty. There have been times when, as I leaned over my work, I could make almost no sense of it. But as I step away now, the scattered details begin to merge into strangely familiar forms. And suddenly it no longer seems random at all, this ending. My plot, I realize, looks just like one of those wayward pages from the Island of Lost Maps.

DETAIL FROM A CELESTIAL CHART
PUBLISHED IN 1515 BY THE ARTIST
ALBRECHT DÜRER.

Lifting Off

*T*HE WORLD WAS BEHIND ME, AND BEFORE ME was nothing but emptiness. What a fine spot it was, this rise overlooking the Pacific, to celebrate the end of my adventure. A crisp spring wind rustled through the low scrub on the hillside; it smelled, as sea air always does, of old tales and new dreams. On the horizon the dull ocean faded haltingly into a leaden sky. For a moment, peering past the outermost limit of the continent, unburdened of all those years of research and obsession at last, I felt almost weightless. Then a voice brought me back. Solemn and scratchy, it crackled from a pair of loudspeakers set up to my rear:

T-MINUS TWO MINUTES . . .

Mapmaking is an endless quest for perspective. From the earliest times human beings have gazed up with awe and envy as the birds spired overhead. What did those soaring creatures see? Where were they going, so far over the horizon? What could be found there, and what lay beyond? How big was the world? Who else lived in it? How was it shaped? Wings not being our evolutionary good fortune, the human mind evolved an ability to dream its way into the skies and peer down, then created a language of symbols to describe what it saw. The first maps came no later than the Stone Age, and the earliest extant map of the world was made about 600 B.C. The Babylonians who crafted that little clay tablet figured the Earth to be flat, but the Greeks realized it was a sphere and came up with a way to project it onto a two-dimensional plane. The Chinese, too, learned how to reduce space to a systematic grid, and threw in another innovation: the magnetic needle. Other cultures, meanwhile, came up with more navigational tools, using the sun and stars to define the Earth. The tools got better. So did the maps. Better maps begat bigger dreams. New continents came into view. Sailing vessels circled the globe. Balloons levitated above it, then airplanes, then rocket ships. Orbiting on one of those ships in 1962, the astronaut John Glenn observed: "I can see the whole state of Florida just laid out like a map." For the first time our perspective had outstripped that of the birds. Even so, we kept yearning for a better view.

I was about to witness the latest chapter in that quest. Gathered around me, hugging themselves against the ocean breeze, were a handful of reporters, space industry executives, and officials from Vandenberg Air Force Base. It was April 1999, and we all had come to this isolated edge of the base, some 170 miles north of Los Angeles, for a pivotal moment in the history of "remote sensing"—the geeky buzzword for examining the Earth from on high. To the south of us, past the Santa Ynez River, loomed a row of hills, behind which, just out of view, stood a ninety-three-foot-tall Athena II rocket, ready to roar into space. Its payload was the most sophisticated Earth-observation satellite ever built for the commercial market. Operated by Space Imaging, a cutting-edge Colorado firm, the *Ikonos I* would be able to produce photographs of objects as small as one meter—images equivalent in detail to maps drawn by the U.S. Geological Survey. From an orbit some 420 miles above the Earth, its camera could distinguish cars, homes, roads, bridges, individual trees, and even hot tubs. Previously the domain of only government spy satellites, these high-resolution images could then be merged with those of another onboard sensor, whose multispectral scans would reveal information not seen with the naked eye, such as the health of crops and the spread of plankton at sea, as well as earthquake fault lines and potential mineral deposits. Transmitted to Earth in digital form—meaning they could be easily stored in computers, then processed and converted into maps—these images would be available almost immediately to anyone willing to pay thirty to three hundred dollars per square mile photographed. Among the early customers was the National Imagery and Mapping Agency, the federal organization responsible for military maps and surveillance photos. But cartographers and spies were far from the only people interested in the *Ikonos I*. At the edge of the millennium, everyone from farmer to miner, urban planner to relief worker, was demanding a bird's-eye view of the world.

Personally, I had a different sort of perspective in mind. In this place where the land and sea and sky converged, where the past and present of cartography met its future, I wanted to take one last look back at a story that had consumed me for four years. Although the case had largely disappeared from public view, it remained very much alive for the people affected by it, many of whom had continued to pursue it behind the scenes. Of course, Bland had never been far from my thoughts, either. I had even had something of a last run-in with him. And so, as one countdown crackled in my ears, another one ticked off in my head, as I thought back on a parade of events that had transpired since the map thief's release from prison.

T-MINUS ONE MINUTE . . .

In February 1998 I had found myself in Florida, where organizers of the Miami International Map Fair were holding a panel discussion on security issues, focusing on the Bland case. Just three years earlier the map thief had passed himself off as a legitimate dealer at the event, trafficking his illicit stock to a number of unknowing map sellers and collectors. This fact was deeply embarrassing for all involved—but, to his credit, Joseph H. Fitzgerald, the fair's affable founder and chairman, chose not to hide from it. Unfortunately, although most of the dealers I met at the fair seemed well-meaning, little was said or done that day to convince me that the same thing could not happen at an event like this again.

Meeting in a basement lecture hall at the Historical Museum of Southern Florida, the panel included two rare books librarians, an art curator, a map seller, and a prominent collector. They covered a broad range of topics, from ways to keep thieves out of libraries to methods for getting maps back once the thieves got in. Noticeably absent from the discussion, however, was an honest examination of how Bland had so easily managed to gain the trust of dealers at this same fair. Such a discussion, however, would have fallen on deaf ears, since almost none of those dealers had bothered to attend the forum. They had better things to do just then. The workshop coincided with the opening of the fair—and in the lobby upstairs, where the dealers had set up their booths, there were customers to meet, deals to cut, maps to move.

"There's a feeding frenzy up there," I heard Fitzgerald tell an associate as he stood near the rostrum, "which is good, but . . ."

He finished his thought by casting a nervous glance at the rows of empty seats.

After the forum was over, I went upstairs and wandered from booth to booth. Dozens of dealers—from such far-flung spots as Argentina, Australia, Canada, England, Germany, and the Netherlands—busily hawked their goods, while hundreds of connoisseurs and other interested people shuffled about, gazing at maps on walls, flipping through maps in bins, holding maps at odd angles to get them in the best light, gathering in groups to discourse about a certain unusual specimen. Most of the fairgoers were middle-aged and older, but there were also twenty- and thirty-somethings, some of whom wore looks of delight and wonder, having just been initiated into the passion's quirky marvels. It was an unlikely event, half subdued art opening, half frenzied swap-o-rama. Maps were changing hands at a brisk pace: one dealer told me he moved $38,000 in product that day. No one seemed outwardly to be concerned about provenance.

Later that morning, away from the hubbub, a veteran dealer approached me in the museum's library. He had a story to tell—one that not even the FBI had heard. I'm not sure why he decided to let me in on it: perhaps he wanted to ease his conscience or perhaps he just figured I'd be interested. Three years earlier he had purchased an item from Gilbert Bland at a fair in San Francisco. He had not tried to get it back to the rightful owner. "The authorities never contacted me," he said with a shrug, "and, well, I'd already sold the map. I got lucky on that one, I guess."

T-MINUS FIFTY SECONDS . . .

"Librarians," said the woman with short red hair, "have an ostrich mentality when it comes to security: they have their heads in the sand and their tails in the air, and they're ripe to be screwed."

Four months had passed since the map fair in Miami, and I was at another workshop devoted to the bad deeds of Gil Bland. This time, the attendees were rare books librarians from around the country, who had gathered in Washington, D.C., for an annual meeting. In a conference room at the Library of Congress, some forty workshop participants had just heard from security experts and law enforcement personnel—including the FBI's Gray Hill—as well as officials from Johns Hopkins, who described Bland's capture and explained their controversial decision not to have him arrested.

Citing the confidential nature of the discussions about security, organizers had asked me not to attend the main body of the workshop. They did, however, invite me to speak at the end of the session, so that I could brief the librarians on my coverage of the case and ask for their input. Not long after stepping off the podium, I was confronted by the woman with the red hair and the vivid ostrich analogy.

Her name was Eileen E. Brady, and it turned out that she was not just blowing off steam. In addition to her work as a librarian at Washington State University, Brady was editor of *Focus on Security: The Magazine of Library, Archive, and Museum Security,* a position that had given her a "healthy paranoia" about the dangers that lurk in the stacks. Theft, she said, was not the only security problem; librarians have been raped and even murdered on the job in recent years. Nonetheless, she reported, many of them would still prefer to ignore the problem. "The attitude that the library is a safe haven, a quiet place, that nothing bad will happen there, that we must not do anything to offend our so-called 'user' or patron—whether or not that patron is legitimate—is very common. And it's difficult to overcome this mind-set.

"My position is militant; there are no two ways about it," she added. But, unlike most militants, Brady was not being shunned by her peers. She had, in fact, been one of the featured speakers at the day's workshop—which struck me as a very healthy sign. Yes, the librarians had a long way to go in defending themselves against the Gilbert Blands of the world. But, unlike the dealers, they seemed sincerely interested in confronting their own demons.

I became even more confident in this conclusion the next day, when I sat down with Everett C. Wilkie, Jr., a sardonic, cigarette-puffing consultant in rare books and manuscripts who had organized the workshop. "What surprised me the most," he told me, "was how cranky the law enforcement people were with the library community in general. They were saying, 'A, you won't prosecute; B, you won't put identification marks on your stuff; C, you don't take care of it; D, if we find it, you won't help us to get it back to you.' They were really very blunt about it."

"Did the librarians stick up for themselves?" I asked.

"No," he said with a glib smile. "It was all true."

Nonetheless, said Wilkie, "I don't think there was anyone in that room who actually wants their stuff stolen, and I think they have a pretty good idea how they would stop it if they only could. But a lot of it—security equipment, staff, you name it—comes down to money. The first thing you've got to do to affect your security program is to get your upper administration to buy into it. And if they don't buy into it, you've got problems."

In fact, I knew that some administrations were beginning to see the light—thanks, ironically, to Bland. His alleged thefts from the University of Florida, for example, served as a much-needed wake-up call to school officials. "Bland did several thousand dollars' worth of damage," said the librarian John E. Ingram, "but the ultimate effect of his 'work' is that we have better security for our stacks, we have better security for our reading room, we have surveillance, which we didn't really have before—and in the final analysis, we have a much more astute, perceptive staff working with the collection, which is probably the best benefit of all." Likewise, major reforms had been put into place at the University of Rochester library, which Bland is thought to have visited three times, once taking so much material that he left the floor snowy with paper shards. After purchasing a camera for the reading room, the rare books department sought the help of a nationally known security consultant in enacting additional measures—which were funded, appropriately enough, by insurance money from the Bland case. "The whole library has much better security, and the rare books department has vastly better security than when Gilbert Bland was here," reported the rare books librarian, Evelyn Walker.

Some observers, however, argued that even more fundamental reforms were necessary. Not the least—and certainly not the least restrained—was the irascible map mogul Graham Arader. "Some of these libraries now have hundreds of millions of dollars of books, and they expect to get security paying people who they get for thirty thousand dollars a year. They're going to have to start treating books like other objects worth as much money," he said. "They're going to have to segregate them and put them in rooms where they're virtually inaccessible. The books are going to have to be stored like gold bullion."

Librarians, he said, could learn a thing or two about security from their corporate counterparts: "The prop room at Disney—I guaran-goddamn-tee you that the curator of Disney knows where everything is and it's on inventory, and he's watching everything. That guy's on top of what he's doing, because if he wasn't, [Disney CEO Michael] Eisner would fire him. One of these librarians who are so critical of me wouldn't last a week there."

Never one to think small, Arader proposed nothing less than the complete elimination of most traditional rare books rooms, in favor of a few centralized and hypersecure research libraries. "There's no reason for any city—even New York—to have more than two major map collections," he said.

When I mentioned this concept to Wilkie, he found it impracticable, pointing out that the larger scale of such facilities might cause as many security problems as it solved. If it was difficult to keep tabs on someone like Bland in a small space, he said, imagine how hard it would be in a huge room filled with hundreds of researchers. Still, Wilkie did not think Arader was completely off base. He noted, for instance, that more and more research libraries are limiting access to serious scholars. To get in, patrons must provide—in advance—a precise description of their research and letters of recommendation. "They are just totally unapologetic about the fact that the general public isn't welcome," he said.

That idea, I told him, struck me as sad.

"It's sad, but it's not sad," he responded. "A lot of these places have stuff that just shouldn't be handled very much. I mean, you really should need to have a good reason to look at this stuff because it's got to last, well, basically forever."

It was hard to argue with that. The legacy of the Bland case, in fact, may be to help turn back the clock on our concept of libraries. In the mid-nineteenth century, when establishing his masterpiece of a research facility in Baltimore, George Peabody had insisted that, while it would contain "the best books on every subject," it would also be "for the free use of all persons who may desire to consult it." It was a progressive notion back then but one that seems to have failed the test of time. In the long run rare books rooms may look less like what Peabody envisioned and more like those of ancient times—repositories for sacred documents available to only a priestly caste, the "Learned Men of the Magic Library," as the Egyptians had called them. And while that system would be sure to have unfortunate consequences—isolating average people even further from their culture and their history—the current one may be even worse. As Wilkie put it: "Any halfway smart undergraduate student can pay his way through four years at a major university for the price of a razor blade."

T-MINUS FORTY SECONDS . . .

In June 1998, just a few days after my conversation with Wilkie, I returned to the Island of Lost Maps. In the more than a year that Gilbert Bland had been out of prison, FBI Special Agent Gray Hill had remained hard at work cleaning up the map thief's mess. Now, at long last, it looked as if Hill's exile was coming to an end.

The agent had just returned from the nation's capital, where, in addition to addressing the rare books librarians, he had met with an FBI technician to match one last set of maps with the books from which they had been taken. "This was my last run up to see the forensics expert," said an obviously weary Hill. "As far as I'm concerned, I'm done at this point."

Since the last time I saw Hill, two years earlier, he had managed to find homes for more than one hundred additional maps. It had been methodical and often frustrating work, but he could take heart in knowing that his efforts had not gone unappreciated. "One of the librarians came in with a real abrasiveness," he recalled. "But she went back with more maps than she had even known she was missing. At the end, she had tears in her eyes."

All told, he had been able to reunite nearly 180 of the roughly 250 wayward maps with their owners. Nonetheless, the Island of Lost Maps still had plenty of inhabitants, most of which were now scattered on a table in front of us at Hill's Charlottesville office. "I feel good about what I've done," he said, "but it does still halfway tick me off that some of them could possibly be matched up. . . . I've got history sitting here, and it should be returned to where it came from."

Did Gilbert Bland harbor the secret of where those remaining maps belonged? Hill found it unlikely. True, Bland had been less than honest with the FBI—failing to tell investigators, for instance, about his visits to libraries in Canada. But Hill believed simple forgetfulness probably also came into play. "When you've stolen from so many places," he said with a faint grin, "it's hard to remember where you got things."

And no matter what information Bland might be withholding, more maps could have been returned if only librarians had taken better care of their materials and kept better records, according to Hill. Fresh from dressing down librarians at the workshop in Washington, he was still full of venom. With some librarians, he said, the failure to take care of materials smacked of pure negligence. "It's like if you had a five-thousand-dollar bill in your wallet, and you said, 'I know I had one in there, but if I look it might be missing. So I just won't look.'"

If the libraries Bland had visited "spent a little more energy" searching for missing items, Hill believed he could get rid of even more of his maps. Still, he was haunted by the possibility that the remaining ones "belong to a library that we are not yet aware of."

In any event, it was no longer his concern. After two and a half long years on the case, he was ready to move on. It had been both an exasperating and an enlightening experience. "I have picked up a much greater appreciation for the art," he said. "The word *cartography* was not a normal part of my vocabulary before this case. So it has been an education, a broadening of the horizons."

"Are you going to miss having them around?" I asked.

He cast one last glance over the pile. "No," he said, looking genuinely relieved.

But, as often happens to those shipwrecked on islands, Agent Hill's dreams of rescue proved premature. When I called him many months later, the maps were still locked in his office, awaiting a permanent home. They would almost certainly wind up at the Library of Congress, Hill said. As usual, however, the federal bureaucracy was moving slowly, and final plans had not yet been completed.

In the meantime, Hill had managed to place a couple more maps not previously known to have been stolen. "Every once in a while somebody will be going through a book," he explained, "and all of a sudden they'll say, 'Hey, you know we're missing something out of here.' . . . It keeps coming in very slowly."

It would continue to come in slowly, this story. Or so I reflected as I stood on that rise, listening to the countdown and watching the waves roll in. The map thief's tale, I thought, might still be unfolding decades from now. I stared off into the lead gray sky. Somewhere beneath the distant clouds, unopened books sat on dusty shelves, crimes waiting to be discovered. And somewhere, too, beautiful maps hung on the walls of perfect homes, crimes waiting to be solved.

T-MINUS THIRTY SECONDS . . .

The last person I ever expected to hear from again was Gilbert Bland—and it's a good bet that the last person he ever planned on sharing his thoughts with was me. Yet, despite it all, there I was one day, furiously taking notes while he spoke about his life.

This final turn of events, like so many of the best discoveries, came out of the blue. It arrived in the form of a videocassette, provided by a source familiar with the case, who had offhandedly asked if I would be interested in watching it.

Somewhat less offhandedly I had said that sure, I might not mind at all. And so, within seconds, Bland appeared before me, neatly attired in jeans and a blue-on-blue-checked flannel shirt, a white T-shirt barely visible beneath his high-buttoned collar. For a moment I just stood there gaping, dumbfounded that years of waiting to hear him tell his story had suddenly ended with nothing more than the press of a PLAY button.

I had heard about this video before. It had been making the rounds among law enforcement types and librarians—shown, for instance, at that workshop on the case in Washington. Conducted by two campus police officers at the University of Delaware, the interview was filmed on March 27, 1997, near the end of Bland's incarceration. As it got under way the map thief sat at a table, sometimes crossing his arms or nervously pressing the fingertips of both hands against his reddish white sideburns, but more often resting his chin on one palm. He looked exhausted and sullen, at one point requesting to be placed in a cell away from other prisoners. "I'm under [psychiatric] medicine," he explained. "And sometimes when I've been on medication, people have beat me up before, and the thing is I literally can't defend myself. I'm not a violent person."

Still, this was not the timid Bland I had seen in court—nor, for that matter, the combative Bland I had heard over the phone. Instead, he seemed intelligent, polite, articulate, and, above all, as slippery as black ice—perhaps not unlike the Bland who had visited those libraries. Evasive about the details of his crimes, he often claimed amnesia:

Officer: And where are [the maps stolen from the uni-
versity] at now?

Bland: I don't know.

Officer: You don't have any idea? Does the FBI have
them?

Bland: The FBI might have them.

Officer: Did you sell any of the maps that you took
from the University of Delaware's books?

Bland: I could have, but I couldn't be sure.

Officer: You couldn't be sure?

Bland: Right. . . .

Officer: Who would you have sold them to?

Bland: I—I don't remember that.

Officer: We're not asking for names. I mean, did you
sell them to art dealers?

Bland: I went to map shows and sold maps at a map
show. I don't know who I sold them to. It's been so
long ago, and I've been through so much in the last
year and a half, that I just can't remember.

At other points, however, his memory seemed to return with
uncanny precision. He had no problem, for instance, in providing a detailed description of the room from which he had stolen
maps. Nor had his methods of operation slipped his mind:

Officer: How did you get them out of the library?

Bland: I folded them up and put them inside my shirt.

Officer: Inside your shirt? Nobody [was] in there to stop
   you from doing it or to see you?

Bland: Well, when they left the room . . . I would just
   fold it up and put it in my shirt and walk out and
   take it with me. . . .

Officer: Put it in your shirt and then leave with it, huh?
   Didn't that destroy the value of the map by folding it
   like that?

Bland: No.

Officer: It didn't?

Bland: No, you'd get home and iron it out, and whatever creases were in it just came out.

Officer: How many other universities did you do this at?

Bland: Uh, a lot.

He was apologetic for his crimes, but the only explanation
he offered was the one given by his lawyers in court: "Business
was really bad and I was about to lose everything I had. And I
have a mental problem that was untreated at the time, and as a
result I came up with the crazy idea to do this. . . . I wish I was
in my right mind. I never would've done this."

The officers, to my frustration, were not particularly interested in his motivations. Their main concern was how future
Gilbert Blands could be stopped. His advice was as simple as it
was self-revealing: "If there would have been a [security] camera there I never would've went in, I never would've stayed
there, I never would've ordered any books . . ."

An invisible man hates cameras the way a vampire hates daylight. The most telling picture I ever saw of Bland shows no face. In that shot, taken outside a courthouse by a photographer for the Raleigh, North Carolina, *News & Observer,* the map thief stands with his cuffed hands held high, not only to obscure his eyes but to obliterate all of his features, as if he were trying by sheer will to make himself disappear. He had almost pulled it off: his crime spree left no film. But here at last he had been captured on video.

And yet, watching his image on that TV screen was strangely like seeing a satellite photograph of the Earth. On the one hand, it gave me a perspective that would otherwise have been impossible. I watched in fascination, for instance, as Bland slid his hand inside his shirt to demonstrate how he stole maps. I had imagined that moment a thousand times, but seeing it on tape gave me a real sense of witnessing the crime. On the other hand, this was a voyeurism of an oddly distant kind. It was as if I was viewing the Planet Bland from hundreds of miles in the air: I could make out the houses and buildings and factories— the facades and end results—but the life that hummed within them, the forces that had created them, remained hidden from view, a world only remotely sensed.

T-MINUS TWENTY SECONDS . . .

The breeze off the Pacific seemed to intensify, as if to keep pace with the collective sense of anticipation. Those of us gathered for the liftoff now squinted anxiously in the direction of the launchpad, knowing that we were about to see history. Not only was the satellite itself a landmark piece of technology—that very morning *The New York Times* had described it as "the world's most powerful civilian spacecraft for observing the Earth"—but its launch came during a revolutionary moment in the history of cartography. As the science writer Stephen S. Hall explained in *Mapping the Next Millennium:* "We find ourselves in the midst of what is arguably the greatest explosion in mapping, and perhaps the greatest reconsideration of 'space' (in every sense of that word), since an anonymous Babylonian first attempted to organize human knowledge of the physical world by drawing a map of the world on a clay tablet twenty-six centuries ago."

As we head into the twenty-first century, the last great topographical mystery on Earth—the global seafloor—is finally surrendering its secrets. In 1995, for instance, scientists armed with freshly declassified spy satellite data produced the first detailed map of this vast terrain, 70 percent of the Earth's surface but virtually uncharted until recent decades. As one researcher explained: "It's like being able to drain the oceans and look at the Earth from space." The heavens, too, are being charted on an almost unimaginable scale. The New Mexico–based Sloan Digital Sky Survey—one of several ambitious space-mapping ventures now under way worldwide—is using a high-tech telescope to chart one quarter of the sky, peering 1.5 billion light-years not only into space but also back in time. By the time it's complete in 2005, the survey will have mapped 100 million galaxies, most of them now unlabeled and unstudied, along with several hundred thousand quasars and tens of millions of individual stars. The undertaking is the first attempt to digitize the universe—to put the heavens in a computer database available to all. "But it's just a first step," said Michael S. Turner, the project spokesman and chairman of the Astrophysics Department at the University of Chicago. "I think it is a very conservative prediction to say that by the end of this century, astronomers will have mapped the entire observable universe." No less staggering are the maps being made of microscopic worlds. As I write these words, scientists are in a frantic race to chart the human genome—the blueprint for each human being, made up of some 100,000 genes. All maps have consequences, but this one is certain to change the course of humanity. By allowing researchers to understand how the body works at its most fundamental level, the genome map could lead to the elimination of a dizzying array of diseases, from muscular dystrophy to cancer and diabetes. It may also result in the prolongation of life, even the retardation of the aging process— Ponce de León's Fountain of Youth discovered at last. And a leading genomist has declared that humankind is on the brink of doing what only God has done so far: create life-forms. Not cloned life-forms—entirely new life-forms, stitched together out of genetic scraps the way Dr. Frankenstein made his monster out of body parts. In the short run, such cut-and-paste organisms would be possible only on a microbial level—but someday the Earth may be prowled by complex beings every bit as "unnatural" as the creatures once found only on the edges of old maps and in the hinterlands of human thought.

Ah, but what of those hinterlands? With all the other stunning gains being made in mapping, will it also become possible to chart the complex terrain of the mind? A growing number of experts believe so. While some scientists study the genome in search of inherited causes for behavioral traits such as alcoholism, aggression, and risk taking—even criminality—others are mapping the brain itself. New imaging techniques allow researchers to look inside the living, working brain, thus "opening up the territory of the mind just as the first oceangoing ships once opened up the globe," in the words of the science writer Rita Carter, author of *Mapping the Mind*. Using state-of-the-art brain scans, scientists are able to pinpoint the exact locations of various brain components—meaning they can literally chart and observe the mechanics of our thoughts, moods, and memories.

These maps promise to inspire some bizarre journeys. The day may come when expectant parents can genetically alter their unborn children to make them less aggressive or impulsive, more cheerful and altruistic. The day, too, may come when all manner of problems, from mental illness and antisocial behavior to post-traumatic stress disorder, can be managed with psychoactive treatments so precise that "an individual's state of mind (and thus behavior) will be almost entirely malleable," according to Carter. The day may come, in short, when to understand the actions and motivations of another person, we must merely examine the peculiar geography of that individual's brain. Nonetheless, my own amateur experiments in the mapping of gray matter lead me to be skeptical. The physical brain might yield itself to this new kind of cartography, but my bet is that the mind—that strange borderland between inner life and outer experience, dreams and memory, body and soul—will prove as difficult to chart as the continent of Atlantis. I have come to believe that humans will always remain fundamentally inexplicable. Then again, there were those who used to say that about the Earth. And, as Carter observed: "These are the early days of mind exploration and the vision of the brain we have now is probably no more complete or accurate than a sixteenth-century map of the world."

T-MINUS TEN . . .

NINE . . .

EIGHT . . .

SEVEN . . .

SIX . . .

FIVE . . .

FOUR . . .

THREE . . .

TWO . . .

ONE . . .

IGNITION . . .

AND . . .

LIFTOFF.

It shot above the hills, a dazzling orange glow. As it rose, everything else seemed to go still; I would have believed the whole world was staring at the sky in wonder just then. It moved slowly and steadily away, a great distant fireball, and then it was gone, leaving behind a vast trail of smoke that snaked up to the heavens like some scribbled message, a cosmic scrawl I could not hope to decipher but felt I had been pondering for a very long time.

The satellite never reached orbit. Minutes after liftoff it went silent, disappeared. They never found it—not in space, not on land, not in the sea. It had simply vanished, swallowed by the great unknown.

Acknowledgments

LIKE A MAP, A BOOK IS THE PRODUCT OF NOT
one person but many. I am profoundly grateful to all those who
opened their doors, lives, and hearts to me as I researched this
project. The librarians Gary L. Menges, Cynthia Requardt,
R. Russell Maylone, John E. Ingram, Evelyn Walker, Alice
Schreyer, Brenda Peterson, and Jean Ashton, among others,
were courageous and candid in discussing a very painful subject.
The map dealer W. Graham Arader III gave me extraordinary
amounts of time and access. Several others in his profession
were also extremely forthcoming, especially Barry Lawrence
Ruderman and the dealers I identified as Once Bitten and Twice
Shy. A number of law enforcement officials, including Donald
Pfouts, Thomas W. Durrer, and Clay Williams, were also vital
to the making of this book. I want to give special thanks to Special Agent Gray Hill, as well as to the FBI offices in Richmond

and Charlottesville, Virginia. In addition, I want to express my sincere appreciation to Jennifer Bryan, Vera Benson, Werner Muensterberger, Harriette Kaley, Hugh Kennedy, Don Etherington, John D. Bergen, Eileen E. Brady, Everett C. Wilkie, Jr., Heather Bland, Selby Kiffer, "Mr. Atlas," and all the others who were kind enough to be interviewed for the project.

This book would not have been possible without three people in particular. Jon Karp, my gifted editor at Random House, took a risk on an unproven author, then guided him through the publishing process with unfailing wisdom and patience. I also feel blessed to have worked with Sloan Harris, a literary agent of extraordinary skill, humor, and honesty. Finally, I owe an incalculable debt to the journalist Michael Paterniti, a driving force behind this project from start to finish. He was a superb critic and mentor—and an even better friend.

I am hugely grateful to Norman J. W. Thrower—a noted cartographic historian and author of *Maps and Civilization: Cartography in Culture and Society*—who read a draft of this text in an effort to ensure its factual accuracy. I was also lucky to get input from a number of other writers and friends. Kevin Davis was there on a daily basis for advice and encouragement, not to mention lunch at the Lincoln Grill. Sara Corbett inspired some breakthrough ideas early on, then helped guide the project through to completion. Bill Lychack, as always, made me dig deeper, think harder, write better. Elizabeth Lychack gave my copy a thorough scrubbing. Clay Harper offered both the pull-no-punches honesty of a trusted old friend and the shrewd perspective of a book industry insider. The awesome Tom Clynes and my talented screenwriting partner, David Freeman, also made invaluable suggestions about the text.

I consider myself blessed to have worked with production editorial guru Benjamin Dreyer and his superb staff: copy editor Susan Brown and proofreaders Evan Stone and Ruthie Epstein. I would also like to thank Amelia Zalcman, Deborah Foley, Andy Carpenter, Barbara Bachman, Richard Elman, Monica Gomez, and Janelle Duryea at Random House, as well as Teri Steinberg, Richard Abate, and Laura Paterson Davies at ICM.

Several editors at *Outside* magazine, including Hal Espen, Mark Bryant, Brad Wetzler, and Katie Arnold, worked on the story that led to this book. I'm especially grateful to Hampton Sides, who gave my raw article much-needed shape and focus.

Some pundits say writing can't be taught: they're wrong. I am eternally indebted to a number of marvelous teachers, both in the classroom and in the newsroom, who have challenged and inspired me over the years. Rita DuChateau, Bob Reid, Charles Sanders, Sheryl Larson, James Weinstein, Ray Elliott, Larry Doyle, Sharon Solwitz, Nicholas Delbanco, and Charles Baxter—thanks. You gave luster to my world as well as to my prose.

Many other people contributed, directly or indirectly, to *The Island of Lost Maps*. Will Tefft offered exceptional support and proposed marvelous schemes. Robert Karrow and the staff of the Newberry Library made me feel welcome at their wonderful cartographic collection. Richard Cohen kept me sane, not to mention out of debt. I also want to thank, in no particular order, Camille Altay and her wonderful clan, Gregg Chaney and Suellen Semekoski, Helena Kozik, Chris Rose, Chris Carr, Jim Beatty, F. J. Manasek, Lloyd and Marie Green, Shari Joffe and Andrew White, Harry Appelman and Mimi Brody, Jason Harper and Janet Siroto, Tom Kosinski, Alan Sanchez, Dolores Riccardo, Merla Mihalik, Wayne Sallee, Jackie Burford, Miles Burford, Tim Gould, Susan Condon, Dave Cullen, Maggie Garb, Cian Gallagher, John Hammond Moore, Elizabeth Schaaf at the Peabody Institute, Fred Musto at Yale, Linda Lidov at Space Imaging, Joseph H. Fitzgerald and Marcia J. Kanner at the Miami International Map Fair, Norman Strasma and Linda Mickle at the International Map Trade Association, George Combs at the Lloyd House, the members of my writers' group, the Cabin Thing crew, El Grupo del Hombres, the Downers Grove gang, *The Daily Illini* and its discontents, and the Tecal-itlán Culinary Army.

My late father, Robert Harvey, gave me a love for both maps and words. This book is his as well as mine. I'm also grateful to my brother, Matthew, for his love, strength, biting humor, and free legal advice. And how does a son even begin to thank his mother? In this case, he says: *I appreciate the interview.* It was typically brave of Tinker Harvey to discuss, for public consumption, one of the darkest chapters of her life. She never ceases to amaze and inspire me.

Finally, to Rengin and Azize—thank you for more than I can express. Writing this book has been wonderful, but building a life with you is a greater joy and accomplishment by far.

Interview

M ILES HARVEY'S DEBUT NONFICTION
work, *The Island of Lost Maps*, was published in September
2000. Since then, the book has become a surprise international bestseller. We asked Harvey to update us on new developments in the case—and in his life.

Have you learned anything about Bland's crime spree since
The Island of Lost Maps was first published?

A few disturbing details have emerged. I had previously believed that Gilbert Bland's operations were confined to North
America, but a British dealer named David Bannister now
claims that Bland was selling merchandise at a map fair in Great
Britain during the mid-1990s. If true, it means that maps stolen
from U.S. and Canadian libraries may now be hanging in the
shops of British dealers or on the walls of British collectors. Of

course, it also opens the unpleasant possibility that the map thief may have paid visits to libraries in England. An American dealer, meanwhile, has alleged that Bland tried to sell him maps *while in prison.* If true, this could mean that the map thief did not turn over to the FBI all the maps in his possession, as he was required to do under the terms of his plea bargain.

After all your years of research, you noted that the only thing you could say about Bland with absolute certainty is that he cherished his anonymity. Has he offered any response to the book?

Well, I haven't heard from him, but I don't imagine he's real thrilled about it. He is a very private man who, unfortunately, committed not only a very public crime but also a crime *against* the public. Those books he mangled belong to all of us. They're our history, our heritage. So, although it makes me uneasy to think that my work might cause him or his family pain, I'm also very confident that the subject merited coverage. *The Island of Lost Maps* is not a particularly political work, but I do hope that it's helped draw attention to the seriousness of library crime. In the past, police, prosecutors, and university administrators have really tended to ignore this important problem.

Is the security situation improving for libraries?

The case itself was indeed a wake-up call for some libraries, such as those at the University of Florida and the University of Rochester. And librarians from several other institutions have informed me that *The Island of Lost Maps* has helped them make the case for better security measures. Still, there's apparently a long way to go. Our libraries continue to be plundered by thieves such as Bland. Michigan State University, for example, recently suffered the loss of several important atlases.

Most of Bland's customers were his fellow map dealers, some of whom suspected him to be a thief but none of whom went to the police. Do you see any signs of reform in this industry?

I'm certainly no expert on the map business, but there are anecdotal reasons to be skeptical. Not long ago, for example, I was talking with a well-known dealer who insisted that my book had actually underestimated the volume of stolen goods trafficked within the map trade. To illustrate his point, he produced some of his own stock, several old sea charts. Each of the charts had a slight horizontal fold across the middle of the page—which made no sense because the folds clearly had been made *after* the maps were removed from a book. "In the business, we sometimes call these 'library folds,' " he said with a knowing look. When I asked him what he meant, he proceeded to pantomime cutting a map from a book, folding it, and stuffing it into a jacket. He conceded that he had purchased the maps knowing full well that someone had probably stolen them at an earlier date. "If I didn't buy them, somebody else would," he said, adding that since records were rarely kept on the provenance of an individual map, there was "no chance for reform" of the practice. Granted, this may be the cynical view of a lone dealer. Nonetheless, the conversation did not fill me with confidence.

Have you been surprised by the response to *The Island of Lost Maps?*

Well, I would not have guessed that so many people would be interested in a book that's basically about libraries. I've also been a bit taken aback that *The Island of Lost Maps* has turned out to be something of a controversial book, with adamant proponents and detractors among readers and critics alike. But I set out to write a cross-genre book: part true-crime story, part history of cartography, part journey of self-discovery. It turned out a bit like one of those hybrid monsters at the edges of old maps. So I suppose I shouldn't be too shocked that not everyone agrees what the beast is or whether it's beautiful.

What's up next for you as a writer?

I'm not sure yet. I have a bunch of book ideas, several of them having to do with people who, like those in *The Island of Lost Maps,* are monomaniacal about a certain subject. Perhaps it says something about how dull my own existence is, but I seem to have an obsession with other people's obsessions.

Notes

Introduction: Strange Waters

ix   "HOUTMAN CORNELIUS" Carl Waldman and Alan Wexler, *Who Was Who in World Exploration* (New York: Facts on File, 1992), p. 327.

x   *mappery* William Shakespeare, *Troilus and Cressida*, act I, sc. 3, line 205.

xi   "black rock" See John Goss, *The Mapmaker's Art: An Illustrated History of Cartography* (Skokie, Ill.: Rand McNally, 1993), p. 95. See also Steven Frimmer, *Neverland: Fabled Places and Fabulous Voyages of History and Legend* (New York: Viking Press, 1976), p. 83.

xii   "No foreign ship" George Masselman, *The Cradle of Colonialism* (New Haven and London: Yale University Press, 1963), p. 62.

xii   the Portuguese quietly discovered Australia See William Eisler, *The Furthest Shore: Images of Terra Australis from the Middle Ages to Captain Cook* (Cambridge: Cambridge University Press, 1995), pp. 17–36.

xiii   Beach, Lucach, and Maletur See ibid., p. 37. See also Raymond H. Ramsay, *No Longer on the Map: Discovering Places That Never Were* (New York: Viking Press, 1972), pp. 36–41.

xiii "The route itself" Masselman, *Cradle of Colonialism*, pp. 62–63.

xiv "the vanguard of the age" Ibid., p. 87.

xiv "speculation in a totally modern fashion" Fernand Braudel, *The Wheels of Commerce*, vol. 2: Civilization and Capitalism: 15th–18th Century, trans. Siân Reynolds (New York: Harper & Row, 1982), p. 100.

xv Secret Atlas See Lloyd A. Brown, *The Story of Maps* (Boston: Little, Brown, 1949), p. 148.

xv most profitable export See Allen M. Sievers, *The Mystical World of Indonesia: Culture and Economic Development in Conflict* (Baltimore: Johns Hopkins University Press, 1974), pp. 89–99. See also Heinrich Eduard Jacob, *Coffee: The Epic of a Commodity*, trans. Eden and Cedar Paul (New York: Viking Press, 1935), pp. 107–121.

xv *koffie . . . kopi* I could find no reference source containing the exact etymology of the Malay-Indonesian word *kopi*. But Cornell University's John U. Wolff, editor of *An Indonesian-English Dictionary*, informed me that *kopi* was "most likely a loan word" from the Dutch *koffie*. This view was generally supported by a number of other scholars I contacted.

xvi "TAMARAC, FLA." Anonymous author, Knight-Ridder/Tribune story, "Police on the Trail of a Map Thief: Florida Antique Shop Owner Is Suspected," *Chicago Tribune* (December 21, 1995): 14.

xvii "Now when I was a little chap . . ." Joseph Conrad, *Heart of Darkness*, ed. Stanley Appelbaum (1902; reprint, Mineola, N.Y.: Dover Publications, 1990), p. 5.

xviii the bizarre case of Stephen Carrie Blumberg See Nicholas A. Basbanes's account in *A Gentle Madness: Bibliophiles, Bibliomanes, and the Eternal Passion for Books* (New York: Henry Holt, 1995), pp. 465–519.

xix the flesh of Egyptian mummies See Annick Le Guérer's *Scent: The Mysterious and Essential Powers of Smell*, trans.

Richard Miller (New York: Turtle Bay Books, 1992), pp. 87–96.

xix   "Alas, poor Egypt!"   Quoted in ibid., p. 90.

CHAPTER ONE: MR. PEABODY AND MR. NOBODY

3   "There is no other library"   Daniel Mark Epstein, "Mr. Peabody and His Athenaeum," *New Criterion,* vol. 14, no. 2 (October 1995): 21–28. All quotations attributed to Epstein in the chapter come from this article—as does some of the historical material, both here and in Chapter Four.

3   "a cathedral of books"   See *Guide to the Archives of the Peabody Institute of the City of Baltimore: 1857–1977,* comp. Elizabeth Schaaf (Baltimore: Archives of the Peabody Institute of Johns Hopkins University, 1987), p. 17.

4   more than $7 million   See Elizabeth Schaaf, "George Peabody: His Life and Legacy, 1795–1869," *Maryland Historical Magazine,* vol. 90, no. 3 (Fall 1995): 269–285. See also Carl Schoettler, "Banker Gave Baltimore a Musical Institute and the Gift of Giving," Baltimore *Sun* (February 16, 1995): 1E.

4   "the most liberal philanthropist"   Quoted in Schaaf, "George Peabody."

4   "Deprived as I was"   Quoted in Franklin Parker, *George Peabody: A Biography,* rev. ed. (Nashville: Vanderbilt University Press, 1995), p. 25.

5   "I hope may become useful"   George Peabody, "Founding Letter," in *Guide to the Archives,* p. 58.

5   "perhaps grander in its original design"   Parker, *George Peabody,* p. 91.

5   "for the free use of all persons"   Peabody, "Founding Letter," p. 59.

5   "the best works on every subject"   Ibid.

5   "to satisfy the researches"   Ibid.

6   "Taking advantage of the delay"   *Guide to the Archives,* pp.

15–17. Although she is listed as the compiler of this pamphlet, Schaaf informed me that she was, in fact, the author of the section quoted.

8   "should be guarded"   Peabody, "Founding Letter," p. 59.

9   a red notebook   A photocopy of this notebook was provided to me by a source who wished to remain anonymous.

9   "like a small candle"   *The Diario of Christopher Columbus's First Voyage to America, 1492–1493*, trans. Oliver Dunn and James E. Kelley, Jr. (Norman: University of Oklahoma Press, 1989), p. 59. The admiral's journal from the first voyage has been missing since 1554. We know its contents thanks to this abstract, made by Bartolomé de Las Casas in the early sixteenth century.

9   "Or was the mysterious light"   John Noble Wilford, *The Mysterious History of Columbus: An Exploration of the Man, the Myth, the Legacy* (New York: Alfred A. Knopf, 1991), p. 130.

14   "Great Washington"   Herman Melville, *Moby-Dick, or, The Whale* (1851; reprint, New York: Modern Library, 1992), p. 222.

CHAPTER TWO: IMAGINARY CREATURES

19   "Demons assaulted ships"   Quoted in Donald S. Johnson, *Phantom Islands of the Atlantic: The Legends of Seven Lands That Never Were*, rev. ed. (New York: Walker, 1996), pp. 28–43.

19   "one-eyed men"   *The Diario of Christopher Columbus's First Voyage to America, 1492–1493*, trans. Oliver Dunn and James E. Kelley, Jr. (Norman: University of Oklahoma Press, 1989), p. 133.

19   "were not as pretty"   Ibid., p. 321.

19   "back and breasts were like a woman's"   Henry Hudson, "A Second Voyage or Employment of Master Henry Hudson," in

Donald S. Johnson, *Charting the Sea of Darkness: The Four Voyages of Henry Hudson* (New York: Kodansha International, 1995), p. 60.

20   "It is easy enough to move"   Steven Frimmer, *Neverland: Fabled Places and Fabulous Voyages of History and Legend* (New York: Viking Press, 1976), p. 18. For a more scholarly work on the same subject, see Percy G. Adams, *Travelers and Travel Liars: 1660–1800* (Berkeley and Los Angeles: University of California Press, 1962).

20   "It is a wonder"   Quoted in Tzvetan Todorov, *The Conquest of America,* trans. Richard Howard (New York: Harper & Row, 1984), p. 21. The passage is from Las Casas's *History of the Indies,* for which no complete English translation is available. I use translations from various authors.

21   "unlikely, even allowing for exaggeration"   Frances Wood, *Did Marco Polo Go to China?* (Boulder: Westview Press, 1996), p. 149.

21   hugely influential   See C.W.R.D. Moseley's introduction to the Penguin edition of *The Travels of Sir John Mandeville* (Harmondsworth, England: Penguin Books, 1983), pp. 9–39.

22   "good Christian men"   Ibid., p. 173.

22   "I, John Mandeville"   Ibid., p. 123.

22   "who live just on the smell"   Ibid., p. 137.

22   "as big as dogs"   Ibid., p. 183.

22   "a kind of fruit"   Ibid., p. 165.

22   "He was an unredeemable fraud"   Stephen Greenblatt, *Marvelous Possessions: The Wonder of the New World* (Chicago: University of Chicago Press, 1991), p. 31.

23   "The abundant identifying marks"   Ibid., pp. 33–34.

24   "I applied for a credit card"   Transcript from *The People of the State of California vs. Gilbert Anthony Bland, aka Jack Arnett, aka Jason Michael Pike,* Municipal Court of the San Diego Judicial District, October 3, 1973.

24   "The Clerk"   Ibid.

25   The defendant was born   Birth records on file with the Marion County, Ind., Health Department.

27   "It must show the area near the treasure"   Franklin W. Dixon, *The Clue in the Embers* (1955; reprint, New York: Grosset & Dunlap, 1972), p. 106.

27   "Wowee! What a treasure!"   Ibid., p. 162.

27   "These treasures are certainly government property!"   Ibid., p. 163.

28   "We don't expect a reward"   Ibid., p. 174.

28   "There is a disturbing darkness"   Frank McLynn, *Robert Louis Stevenson: A Biography* (New York: Random House, 1993), p. 199.

29   "The doctor opened the seals"   Robert Louis Stevenson, *Treasure Island* (1883; reprint, Cleveland: World Publishing, 1946), p. 62.

29   "monstrous imposter"   Ibid., p. 278.

30   "prodigious villain"   Ibid.

30   "bland, polite, obsequious seaman"   Ibid., p. 279.

30   "figure whose personality swings"   Ian Bell, *Dreams of Exile: Robert Louis Stevenson, a Biography* (New York: Henry Holt, 1992), pp. 150–151.

30   "It was a strange collection"   Stevenson, *Treasure Island*, p. 281.

31   "two old, dirty, and ragged charts"   James Fenimore Cooper, *The Sea Lions, or, The Lost Sealers* (1849; reprint, New York: G. P. Putnam's Sons, 1896), p. 50.

31   "very considerable amount of treasure"   Ibid., p. 35.

32   "no dangers, no toil"   Ibid., p. 314.

32   "I'm afraid that"   Ibid., p. 410.

32   "half-way belief"   Ibid., p. 381.

32   "a little more than two thousand dollars"   Ibid., p. 454.

32   "Seeing the impossibility of restoring the gold"   Ibid.

33   "the greatest literary mystery"   Paul Theroux, "Solving the Traven Mystery," *Sunday Times* of London (June 22, 1980): 44.

33   "Here was someone"   Karl S. Guthke, *B. Traven: The Life Be-*

hind the Legends, trans. Robert C. Sprung (Brooklyn: Lawrence Hill Books, 1991), p. 12.

33   "One of the maps"   B. Traven, *The Treasure of the Sierra Madre* (1935; reprint, New York: Signet, 1968), p. 68.

34   "that eternal curse on gold"   Ibid., p. 76.

37   Maps spoke to me   Charles Baxter's superb essay "Talking Forks: Fiction and the Inner Life of Objects" helped me to clarify my thoughts on this subject. See Baxter, *Burning Down the House: Essays on Fiction* (St. Paul, Minn.: Graywolf Press, 1997), pp. 79–108.

37   "It is almost as if one had to read"   Arthur H. Robinson and Barbara Bartz Petchenik, *The Nature of Maps: Essays Toward Understanding Maps and Mapping* (Chicago: University of Chicago Press, 1976), p. 45.

38   "infinite, eloquent suggestion"   Robert Louis Stevenson, "My First Book," in *The Lantern-Bearers and Other Essays,* ed. Jeremy Treglown (New York: Farrar, Straus and Giroux, 1988), p. 283. All further quotations from Stevenson in the chapter come from this short essay.

39   "napkin of the world"   See David Woodward, "Medieval Mappaemundi," in *The History of Cartography,* vol. 1: Cartography in Prehistoric, Ancient, and Medieval Europe and the Mediterranean, ed. J. B. Harley and David Woodward (Chicago: University of Chicago Press, 1987–), p. 287.

41   the various monstrous races   See chart in ibid., p. 331.

41   "Like the earth of a hundred years ago"   Aldous Huxley, *Heaven and Hell* (London: Chatto & Windus, 1956), p. 9.

41   "the antipodes of the mind"   Ibid., p. 16.

42   "strange psychological creatures"   Ibid., p. 11.

42   "an Old World of personal consciousness"   Ibid., p. 10.

42   "the mind's far continents"   Ibid., p. 84.

Chapter Three: The Map Mogul

51   "pitched fine art"   Hugh Kennedy, *Original Color* (New York: Nan A. Talese, 1996), p. 177.

55   "These Old Maps"   Karen Hube, "These Old Maps Offer You a New Way to Double Your Money," *Money,* vol. 26, no. 3 (March 1997): 172.

59   the Antiquarian Booksellers' Association of America drummed him   See Mark Singer's "Profiles: Wall Power," in *New Yorker,* vol. 63, no. 41 (November 30, 1987): 44–46ff. Some of the other information used in this chapter also comes from Singer's excellent article.

60   "I said, 'Graham, what are you doing?' "   Quoted in ibid.

61   reportedly sent a check for $8,000   See Associated Press story, "Rare Map of Houston, Missing Since '88, Is Returned," *Dallas Morning News* (June 4, 1996): 16D.

64   If you wanted to buy this volume in 1884   See chart in R. A. Skelton, *Maps: A Historical Survey of Their Study and Collecting* (Chicago: University of Chicago Press, 1972), p. 59. In addition to the prices mentioned here, the 1984 figure was supplied by Sotheby's.

68   "More than any one of the ancients"   Lloyd A. Brown, *The Story of Maps* (Boston: Little, Brown, 1949), p. 61.

69   "grinding persistence"   John Hale, *The Civilization of Europe in the Renaissance* (New York: Atheneum, 1994), pp. 193, 190.

69   "Please, if you love me"   Quoted in Nicholas A. Basbanes, *A Gentle Madness: Bibliophiles, Bibliomanes, and the Eternal Passion for Books* (New York: Henry Holt, 1995), p. 72.

70   "insatiable desire"   Ibid.

70   "rare, valuable, or merely strange objects"   Hale, *Civilization of Europe,* p. 530.

70   "caused an immediate and enormous stir"   Thomas Goldstein, *Dawn of Modern Science: From the Ancient Greeks to the Renaissance* (Boston: Houghton Mifflin, 1980), p. 22.

71   "Apart from the extravagant Bibles"   Lisa Jardine, *Worldly*

Goods: A New History of the Renaissance (New York: Nan A. Talese, 1996), p. 205.

71   "The revival of Ptolemy"   Daniel J. Boorstin, *The Discoverers: A History of Man's Search to Know His World and Himself* (New York: Random House, 1983), p. 152.

72   "toward the regions of India"   Passport issued by Spanish sovereigns, quoted in Samuel Eliot Morison, *The European Discovery of America: The Southern Voyages*, A.D. *1492–1616* (New York: Oxford University Press, 1974), p. 43.

73   "were not housed according to their worth"   Quoted in John Addington Symonds, *Renaissance in Italy: The Revival of Learning,* vol. 2 (New York: Charles Scribner's Sons, 1907), p. 99.

73   "For him that steals"   Quoted in Alberto Manguel, *A History of Reading* (New York: Viking Press, 1996), p. 244.

74   Christie's conceded the volume had been stolen   See Susannah Herbert, "International: 1477 Ptolemy Atlas Stolen from Library," *Daily Telegraph* of London (November 7, 1997), as well as an unsigned Reuters story, "Stolen Rare Book by Ptolemy Recovered in London" (May 29, 1998).

76   nearly $800,000   This figure includes buyer's premiums and taxes.

77   computer financial crimes   See Douglas Pasternak and Bruce B. Auster, "Terrorism at the Touch of a Keyboard," *U.S. News & World Report,* vol. 125, no. 2 (July 13, 1998): 37.

Chapter Four: An Approaching Storm

80   "The Spanish sailors"   Daniel J. Boorstin, *The Discoverers: A History of Man's Search to Know His World and Himself* (New York: Random House, 1983), p. 217.

82   "just wanted them"   Cynthia Requardt, text from briefing given to fellow rare books librarians during a panel on the Bland case, June 23, 1998.

82   Murders had risen nearly 9 percent   See Peter Hermann, "Vi-

olent Crime in Md. Rising," Baltimore *Sun* (December 30, 1995): 1B.

83   "We were advised"   Quoted in Frank D. Roylance and David Folkenflik, "Thief Cutting Rare Book at Peabody Was Let Go; FBI Seeks Man Linked Later to Similar Acts at Eight University Libraries," Baltimore *Sun* (December 14, 1995): 1A.

86   "Our goal"   Quoted in John Dorsey, "Auction of Peabody Books Draws Approval, Outrage," Baltimore *Sun* (May 28, 1989): 1F. The quotations from John Burgan, Regina Soria, and Arthur Gutman also come from this article.

90   "It put Baltimore on the world map"   Frank R. Shivers, Jr., "Visit Mount Vernon" brochure, sponsored by the Charles Street Association and Mount Vernon Cultural District.

90   "national treasure"   Brian Lewbart, spokesman for Downtown Partnership, quoted in Paul W. Valentine's "Washington Monument Reopened in Baltimore; Structure Has Been Closed Eight Years for Repair," *Washington Post* (December 5, 1992): C4.

91   upwards of $30,000   An exact price on the Mills atlas is difficult to pin down, since the book is so rare that it does not often change hands. This estimate was made by John Duncan, a retired history professor and owner of V. & J. Duncan Antique Maps and Prints in Savannah, Georgia.

93   "On December 7"   Posting on ExLibris, December 7, 1995.

CHAPTER FIVE: HOW TO MAKE A MAP, HOW TO TAKE A MAP

95   *"others have seen"*   Denis Wood with John Fels, *The Power of Maps* (New York: Guilford Press, 1992), p. 7.

95   "Progress in the science of cartography"   Lloyd A. Brown, *The Story of Maps* (Boston: Little, Brown, 1949), p. 18.

97   "world describers"   See Svetlana Alpers, "The Mapping Impulse in Dutch Art," in *Art and Cartography: Six Historical Es-*

says, ed. David Woodward (Chicago: University of Chicago Press, 1987), p. 59.

98   "maps enable us"   Quoted in ibid., p. 90.

98   copper plate for printing   Most of the material in this section is based on Coolie Verner's "Copperplate Printing," in *Five Centuries of Map Printing,* ed. David Woodward (Chicago: University of Chicago Press, 1975), pp. 51–75.

100   how long it took   See C. Koeman, *Joan Blaeu and His Grand Atlas: Introduction to the Facsimile Edition of Le Grand Atlas, 1663* (Amsterdam: Theatrum Orbis Terrarum, 1970), pp. 43–46.

100   "The planning involved"   Ibid., p. 43.

102   "The crossing of the border"   Avner Falk, "Border Symbolism," in *Maps from the Mind: Readings in Psychogeography,* ed. Howard F. Stein and William G. Niederland (Norman: University of Oklahoma Press, 1989), p. 145.

102   "The theological collection"   Michael H. Harris, *History of Libraries in the Western World,* 4th ed. (Lanham, Md.: Scarecrow Press, 1995), p. 8.

102   God would smite dead   2 Samuel 6:7.

103   "Some people seem to seek"   G. Raymond Babineau, "The Compulsive Border Crosser," *Psychiatry,* vol. 35 (August 1972): 281–290. Further quotations from Babineau also come from this article.

104   "Perhaps the most outstanding conclusion"   *The Warren Report: The Official Report on the Assassination of President John F. Kennedy,* Associated Press, ed. (N.P.: Associated Press, 1964), pp. 160–161.

104   "Countless governments"   Mark Monmonier, *Drawing the Line: Tales of Maps and Cartocontroversy* (New York: Henry Holt, 1995), p. 146.

Chapter Six: The Invisible Crime Spree

107   he introduced the first modern ventilation system   See James
       H. Bready, "Book and Authors," Baltimore *Sun* (July 17, 1949).

108   "If you get bitten by a flea"   Quoted in Peter Young, "Lloyd
       to Assemble Historical Annapolis Data," Baltimore *Evening Sun*
       (November 30, 1960).

108   He delighted in recounting   See Bready, "Books and Au-
       thors." The rough estimate of the number of books listed in
       Lloyd's bibliography is my own.

112   nineteen libraries   In addition to the Peabody Library in Bal-
       timore, the FBI reported that Bland stole materials from the
       following institutions: the American Antiquarian Society in
       Worcester, Massachusetts; the British Columbia Archives in
       Victoria; Brown University; Duke University; the Library of
       Virginia in Richmond; the New York State Library in Albany;
       Northwestern University; the University of British Columbia
       in Vancouver; the University of Chicago; the University of
       Delaware; the University of Florida; the University of North
       Carolina at Chapel Hill; the University of Rochester; the Uni-
       versity of South Carolina; the University of Virginia; the Uni-
       versity of Washington in Seattle; Washington University in St.
       Louis; and Wesleyan University in Middletown, Connecticut.
       In addition, officials at the Newberry Library in Chicago told
       me that someone identifying himself as James Perry visited
       their reading rooms in 1994 but does not appear to have stolen
       anything—probably because the maps he examined had been
       stamped as property of the institution.

113   Learned Men of the Magic Library   See Michael H. Harris,
       History of Libraries in the Western World, 4th ed. (Lanham, Md.:
       Scarecrow Press, 1995), pp. 17–35. See also Alberto Manguel, *A
       History of Reading* (New York: Viking Press, 1996), pp. 187–199.

113   "war with the forces of oblivion"   Umberto Eco, *The Name of*

the Rose, trans. William Weaver (New York: Harcourt Brace Jo-vanovich, 1983), p. 38.

114 "books are humanity in print" Barbara W. Tuchman, "Ex-cerpts of Barbara Tuchman's Lecture at Library of Congress," *Authors Guild Bulletin* (November–December 1979): 15–18.

125 returned by law enforcement officials In addition to the Ogilby maps, two plates from a Mercator atlas were stolen from the University of Washington. Both were later recovered and returned by the FBI.

127 "poets and prose-writers" Quoted in Luciano Canfora, *The Vanished Library,* trans. Martin Ryle (Berkeley and Los Angeles: University of California Press, 1989), p. 20.

128 "it is possible to sail" Strabo, *The Geography of Strabo,* trans. Horace Leonard Jones, vol. 1 (Cambridge, Mass.: Harvard University Press, 1917), p. 433.

129 "Eratosthenes' measurement" Norman J. W. Thrower, *Maps and Civilization: Cartography in Culture and Society,* 2d ed. (Chicago: University of Chicago Press, 1999), p. 21.

129 "distance ruled" Hannah Arendt, *The Human Condition* (Chicago: University of Chicago Press, 1958), p. 250.

129 only 2,760 miles See chart in Samuel Eliot Morison, *Admiral of the Ocean Sea: A Life of Christopher Columbus* (Boston: Little, Brown, 1942), p. 68. Although Morison gives the figures in nautical miles, I have, for the sake of clarity, approximated them here in statute miles.

129 "had not the remotest idea" Samuel Eliot Morison, *The European Discovery of America: The Southern Voyages,* A.D. 1492–1616 (New York: Oxford University Press, 1974), p. 336.

129 "the shrinkage of space" Arendt, *Human Condition,* p. 250.

130 "will help to shrink the world" Frances Cairncross, *The Death of Distance: How the Communications Revolution Will Change Our Lives* (Boston: Harvard Business School Press, 1977), p. 279.

130 "In this modern age, very little remains" Quoted in Peter

Whitfield, *New Found Lands: Maps in the History of Exploration* (New York: Routledge, 1998), p. 186.

131 "must thrill for the saddle" Theodore Roosevelt, *A Book-Lover's Holidays in the Open* (New York: Charles Scribner's Sons, 1916), p. vii.

CHAPTER SEVEN: A BRIEF HISTORY OF CARTOGRAPHIC CRIME

140 The man in the picture See John Goss, *The Mapmaker's Art: An Illustrated History of Cartography* (Skokie, Ill.: Rand McNally, 1993), pp. 141–148.

142 All cultures are thought to make maps See David Stea, James M. Blaut, and Jennifer Stephens, "Mapping as a Cultural Universal," in *The Construction of Cognitive Maps,* ed. Juval Portugali (Boston: Kluwer, 1996), pp. 1–16.

142 "In the Roman Empire" Daniel J. Boorstin, *The Discoverers: A History of Man's Search to Know His World and Himself* (New York: Random House, 1983), p. 269.

142 locked in the innermost vaults See Lloyd A. Brown, *The Story of Maps* (Boston: Little, Brown, 1949), pp. 8–9.

143 having failed to sell Portugal's King John II In fact, the king may have been up to some thievery himself—with Columbus as his victim. Bartolomé de Las Casas wrote that "the King of Portugal wormed more and more information out of Christopher Columbus and [then] . . . he secretly equipped a caravel with Portuguese sailors and set it on the ocean to follow the route Columbus had charted for himself." A storm, however, forced the caravel back to Lisbon. Realizing he had "been the object of a double deal," Columbus "decided to leave Lisbon and come to Castile," wrote Las Casas. See Bartolomé de Las Casas, *History of the Indies,* trans. and ed. Andree Collard (New York: Harper & Row, 1971), p. 23.

143 "fearing the king would send" Quoted in John Noble Wil-

ford, *The Mysterious History of Columbus: An Exploration of the Man, the Myth, the Legacy* (New York: Alfred A. Knopf, 1991), p. 84.

143   "Bartholomeo prepared to join"   Lisa Jardine, *Worldly Goods: A New History of the Renaissance* (New York: Nan A. Talese, 1996), pp. 300–301.

145   the Italian secret agent   Ibid., pp. 107–108.

146   "brought with him a well-painted globe"   Quoted in Samuel Eliot Morison, *The European Discovery of America: The Southern Voyages, A.D. 1492–1616* (New York: Oxford University Press, 1974), p. 319.

146   "In making the globe"   Jardine, *Worldly Goods*, p. 299.

146   Jorge Reinel   See Edouard Roditi's discussion of the Behaim-Reinel question in *Magellan of the Pacific* (London: Faber and Faber, 1972), pp. 103–111 and 129–136. See also Simon Berthon and Andrew Robinson, *The Shape of the World* (London: George Philip, 1991), p. 78.

147   "He knew where to sail"   Quoted in Morison, *European Discovery*, p. 381.

148   "The Spanish . . . kept their official charts"   Boorstin, *Discoverers*, pp. 267–268.

148   "master thief of the unknown world"   Quoted in John H. Parry, "Drake and the World Encompassed," in *Sir Francis Drake and the Famous Voyage, 1577–1580: Essays Commemorating the Quadricentennial of Drake's Circumnavigation of the Earth*, ed. Norman J. W. Thrower (Berkeley and Los Angeles: University of California Press, 1984), p. 2.

149   "prized these greatly"   Quoted in Alexander McKee, *The Queen's Corsair: Drake's Journey of Circumnavigation, 1577–1580* (New York: Stein & Day, 1978), p. 216.

149   "exactly what he required"   Ibid.

149   "Drake's smooth passage"   John Hampden, Introduction to *Francis Drake: Privateer,* ed. John Hampden (Tuscaloosa: University of Alabama Press, 1972), p. 16.

149   "In the sixteenth century"   Brown, *Story of Maps,* p. 9.

150   "a great Book full of Sea-Charts"   Quoted in *A Buccaneer's Atlas: Basil Ringrose's South Sea Waggoner,* ed. Derek Howse and Norman J. W. Thrower (Berkeley and Los Angeles: University of California Press, 1992), p. 22.

150   "The Spaniards cried"   Quoted in ibid., p. 22.

150   "royal influence"   Ibid., p. 1.

151   "the plans of some fortified places"   Quoted in Peter Barber, "Espionage!" in *Tales from the Map Room: Fact and Fiction About Maps and Their Makers,* ed. Peter Barber and Christopher Board (London: BBC Books, 1993), pp. 90–91.

151   the Battle of the Brandywine   See G.J.A. O'Toole, *Honorable Treachery: A History of U.S. Intelligence, Espionage, and Covert Action from the American Revolution to the CIA* (New York: Atlantic Monthly Press, 1991), p. 42.

151   "Spies directed by Washington"   J. B. Harley, "The Map User in the Revolution," in *Mapping the American Revolutionary War,* ed. J. B. Harley, Barbara Bartz Petchenik, and Lawrence W. Towner (Chicago: University of Chicago Press, 1978), p. 104.

151   "men were . . . prepared to risk their lives"   Ibid., p. 105.

151   "brought in with me"   Quoted in ibid., p. 103.

152   "knew no more about the [local] topography"   Quoted in Christopher Nelson, *Mapping the Civil War: Featuring Rare Maps from the Library of Congress* (Washington, D.C.: Starwood, 1992), p. 9.

152   "I *know* that the principal northern papers"   Quoted in O'Toole, *Honorable Treachery,* p. 132.

153   "There was no doubt"   Jay Robert Nash, *Spies: A Narrative Encyclopedia of Dirty Tricks and Double Dealing from Biblical Times to the Present* (New York: M. Evans, 1997), p. 251.

153   "Securing the blueprint"   Quoted in ibid., p. 54.

154   the largest and most spectacular escape   See John Hammond Moore's fine book on this episode, *The Faustball Tunnel: Ger-*

man POWs in America and Their Great Escape (New York: Random House, 1978).

155   "stood traditional cartography on its ear"   Stephen S. Hall, *Mapping the Next Millennium: The Discovery of New Geographies* (New York: Random House, 1992), p. 67.

157   no substitute for a good map   Some subsequent news reports suggested that the embassy bombing was no accident, after all. Quoting "senior U.S. and European military sources," the London *Observer* declared that the embassy was purposely targeted because it had been serving as a rebroadcast station for the Yugoslavian army. As one source at the U.S. National Imagery and Mapping Agency told the *Observer,* the "wrong map" story is "a damned lie." See John Sweeney, Jens Holsoe, and Ed Vulliamy, "Revealed: NATO Bombed Chinese Deliberately," London *Observer* (October 17, 1999): I. See also Joel Bleifuss, "A Tragic Mistake?," *In These Times,* vol. 24, no. I (December 12, 1999): 2–3.

157   a Michigan man named Bill Stewart   See Jack Lessenberry, "Making a Federal Case of It," *Detroit Metro Times* (August 9–15, 1995). A number of people protested Stewart's conviction, arguing that while the material he attempted to sell was technically restricted, it was neither secret nor particularly sensitive. As Lessenberry observed, *National Geographic* magazine had just published "even better maps of the same area."

157   a couple of priests   See unsigned article, "Thieves Steal Valuable Books from Yale Library," *Yale Alumni Magazine,* vol. 36, no. 8 (May 1973): 30. See also John Mongillo, Jr., and Jack Millea, "Two 'Priests' Are Indicted in Theft of Yale Volumes," *New Haven Register* (March 17, 1973): 1–2, as well as Dan Collins and John Mongillo, Jr., "Two Hundred Rare Volumes Missing at Yale," *New Haven Register* (March 18, 1973): 1–2A. I also consulted court records for this section.

158   "Avoid, as you would the plague"   St. Jerome, "Letter 52," *Se-*

lect Letters of St. Jerome, trans. F. A. Wright, eds. T. E. Page,
E. Capps, and W.H.D. Rouse (New York: G. P. Putnam's Sons,
1933), p. 201.

158   unusual rogues' gallery   My list of modern-day cartographic
criminals is not comprehensive, of course. Among the omis-
sions are two notorious U.S. library crooks, Stephen Carrie
Blumberg and James Shinn, neither of whom specialized in
maps per se but both of whom had a fondness for travel literature. I've also excluded European thieves like Ian Hart, who
during the early 1980s stole a fortune in old maps and atlases
from the Bodleian Library at Oxford by hiding them in his
trousers.

158   "was regarded with that special undergraduate awe"   Walt
Philbin, "American Dream Now a Nightmare for Tulane Prof,"
New Orleans *States-Item* (January 13, 1979): A-6.

159   "I was there, I was tempted"   Ibid.

159   "I indeed ruined my life and career"   Quoted in unsigned
story, "Tulane Prof Sentenced to One Year," New Orleans
States-Item (January 8, 1979).

159   "the standard reference"   Unsigned, "Trade Reviews," *AB
Bookman's Weekly,* vol. 85, no. 24 (June 11, 1990): 2524–2525.

159   "few great counterfeiters"   Lynn Glaser, *Counterfeiting in
America: The History of an American Way to Wealth* (New York:
Clarkson N. Potter, 1968), p. 118.

160   In July 1974 Glaser was arrested   See Richard Yurko, "Library
Theft Disclosed; Prime Suspect Arraigned," *The Dartmouth*
(July 9, 1974): I, 6. In preparing this section, I also consulted an
unsigned Associated Press report, "Antique Document Dealer
Is Sentenced for Thefts of Rare Maps from 'U'," in the Min-
neapolis *Tribune* (November 25, 1982), as well as court documents.

160     $300,000 and $100,000   In May 1999, books containing the
1613 and 1632 maps were auctioned for $398,500 and $134,500,
respectively. Selby Kiffer of the rare books and manuscripts de-

partment at Sotheby's informed me that the individual Champlain maps accounted for the overwhelming part of each book's value.

161   "For twenty-five years I have been handling"   Lynn Glaser, *America on Paper* (Philadelphia: Associated Antiquaries, 1989), p. 3.

161   wearing surgical gloves   See Beverly Goldberg, "Judge to Reconsider Map Thief's Probation Sentence," *American Libraries,* vol. 23, no. 6 (June 1992): 429–431.

161   "Mr. Straight"   See Joe Earle, "Ex-UGA Librarian Described as 'Mr. Straight,' " *Atlanta Journal and Constitution* (January 21, 1987): B-1. I also consulted a number of other *Atlanta Journal and Constitution* and *Athens Daily News / Banner-Herald* stories for this section, as well as Nicholas A. Basbanes's account in *A Gentle Madness: Bibliophiles, Bibliomanes, and the Eternal Passion for Books* (New York: Henry Holt, 1995), pp. 488–490.

162   as much as $600,000   This estimate was provided by Selby Kiffer at Sotheby's.

162   "a disgrace"   Quoted in Associated Press story, "Texas Legislature Considers King Holiday" (May 2, 1985). See also David Streitfeld, "Dealer Held in Library of Congress Theft," *Washington Post* (March 13, 1992): F2; as well as Streitfeld, "Book Thief Sentenced to Six Months," *Washington Post* (October 1, 1992): C4.

163   "a rising star"   Ric Kahn, "Two Faces of Serial Thief Bared in N.H.," *Boston Globe* (January 16, 1998): A1.

163   "He only feels comfortable"   Defense attorney Stephen Jeffco, quoted in ibid. I also consulted a number of other *Boston Globe, Concord Monitor,* and Manchester *Union-Leader* stories for this section, as well as court and law-enforcement records.

165   "suspected of delivering weaponry"   Randy Ellis and John Parker, "Man in Dutch Prison Not Tied to City Blast, Authorities Say," *Oklahoman* (December 30, 1995).

165   "I was involved in it"   Quoted in Associated Press story,   .

"American Held in Netherlands Denies Bombing Link," in *Oklahoman* (January 12, 1996).

165 "Thirty Questions About Oklahoma City" See Big Sky Patriot website, June 5, 1997.

167 Johnny Jenkins "Rare Book and Manuscript Thefts," *AB Bookman's Weekly*, vol. 69, no. 7 (February 15, 1982): 1224–1239.

168 "I suppose he was a man" Calvin Trillin, "American Chronicles: Knowing Johnny Jenkins," *New Yorker*, vol. 65, no. 37 (October 30, 1989): 79–97.

168 "[He and his assistant] have orders" Quoted in Berthon and Robinson, *Shape of the World*, p. 79.

172 "failed to mention" Katherine S. Van Eerde, *John Ogilby and the Taste of His Times* (Folkestone, England: Wm. Dawson & Sons, 1976), p. 108.

172 "made a special plea" Brown, *Story of Maps*, p. 169.

173 "made ungenerous use" Letter from Jefferson to Humboldt, December 6, 1813, in *The Journals of Zebulon Montgomery Pike: With Letters and Related Documents*, vol. 2, ed. Donald Jackson (Norman: University of Oklahoma Press, 1966), p. 387. Interestingly, Jefferson made no such excuses for the British cartographer Aaron Arrowsmith, who apparently also borrowed from Humboldt. "That . . . Arrowsmith should have stolen your map of Mexico," Jefferson wrote, "was in the piratical spirit of the country."

174 "Grand Peak" Pike, *Journals*, vol. 1, p. 350.

174 "the highest altitude ever attained" Douglas Botting, *Humboldt and the Cosmos* (New York: Harper & Row, 1973), p. 155. Although he climbed to a record-breaking height of 18,893 feet above sea level, Humboldt stopped some 1,800 feet short of Mount Chimborazo's summit.

174 "The study of maps" Quoted in Helmut de Terra's *Humboldt: The Life and Times of Alexander von Humboldt, 1769–1859* (New York: Alfred A. Knopf, 1955), p. 17.

174 "a truly magnificent cartographic achievement" Carl I.

Wheat, *Mapping the Transmississippi West, 1540–1861*, vol. 1:
The Spanish Entrada to the Louisiana Purchase, 1540–1804
(San Francisco: Institute of Historical Cartography, 1957),
p. 132.

174   "a great variety of data"   Quoted in ibid., p. 134.

177   firsthand information gathered by French explorers   This is
the same Delisle map mentioned in Chapter Ten.

180   *"haunted by the feeling"*   Franz Kafka, *The Castle,* trans. Willa
and Edwin Muir (New York: Schocken Books, 1954), p. 54.

Chapter Eight: Pathfinding

183   "singularly unfavorable to travel"   John Charles Frémont, *Report of the Exploring Expedition to the Rocky Mountains* (1845;
reprint, Ann Arbor, Mich.: University Microfilms, 1966), p. 213.

184   "did little of importance"   Bernard DeVoto, *The Year of Decision: 1846* (Boston: Houghton Mifflin, 1942), p. 39.

184   "The map . . . radically and permanently altered"   Carl I.
Wheat, *Mapping the Transmississippi West, 1540–1861*, vol. 2:
From Lewis and Clark to Frémont, 1804–1845 (San Francisco:
Institute of Historical Cartography, 1958), p. 199.

184   "fixing on a small sheet"   Quoted in Seymour I. Schwartz
and Ralph E. Ehrenberg, *The Mapping of America* (New York:
Harry N. Abrams, 1980), p. 262.

186   At 10:55 P.M.   Police records, Bayonne, N.J.

186   the boy's birth certificate   Birth records on file with the Marion County, Ind., Health Department.

187   "the allegations of the complaint"   Marion Superior Court
records from April 14, 1954.

187   "was both verbally and physically abusive"   Paul R. Thomson, Jr., and James R. Creekmore, "Defendant's Sentencing
Memorandum," *United States of America vs. Gilbert Joseph Bland,*
U.S. District Court, Charlottesville, Va., October 7, 1996.

188   his personal effects   Police records, Bayonne, N.J.

191   "loftiest peak of the Rocky Mountains"   Frémont, *Report of the Expedition to the Rockies,* p. 70.

191   "stupid" and "foolhardy"   Charles Preuss, *Exploring with Frémont: The Private Diaries of Charles Preuss, Cartographer for John C. Frémont on His First, Second, and Fourth Expeditions to the Far West,* trans. and ed. Erwin G. and Elisabeth K. Gudde (Norman: University of Oklahoma Press, 1958), p. 55.

192   "dancing master"   John Moring, *Men with Sand: Great Explorers of the North American West* (Helena, Mont.: TwoDot, 1998), p. 166.

192   "occasional upholstering"   Andrew Rolle, *John Charles Frémont: Character as Destiny* (Norman: University of Oklahoma Press, 1991), p. 1.

192   "No one ever knew why"   Moring, *Men with Sand,* p. 166.

192   "psychiatric techniques"   Rolle, *John Charles Frémont,* p. xiv.

192   "fragmented sense of selfhood"   Ibid., p. 281.

192   "Because Frémont's formative years"   Ibid.

192   "In fact, even while on his expeditions"   Ibid., p. 278.

193   "reckless impetuosity"   Allan Nevins, *Frémont: The West's Greatest Adventurer,* vol. 1 (New York: Harper & Brothers, 1928), p. 114.

194   "To go back was impossible"   Frémont, *Report of the Expedition to the Rockies,* pp. 74–75.

195   "flushed with success, and familiar with the danger"   Ibid., p. 75.

196   "Bereft of two nurturing parents"   Rolle, *John Charles Frémont,* p. 281.

197   "habitual irregularity and incorrigible negligence"   Quoted in ibid., p. 9.

197   "mutiny, disobedience, and conduct prejudicial"   Quoted in Moring, *Men with Sand,* p. 179.

197   "of strange scenes and occurrences"   John Charles Frémont, *Narratives of Exploration and Adventure,* ed. Allan Nevins (New York: Longmans, Green, 1956), p. 29.

197   "There was a map of Vietnam"   Michael Herr, *Dispatches* (New York: Alfred A. Knopf, 1977), p. 3.

199   "has a pattern of problems"   Transcript of sentencing hearing in *United States of America vs. Gilbert Joseph Bland,* U.S. District Court, Charlottesville, Va., December 9, 1996.

201   massage parlor for American soldiers   See Frances FitzGerald, *Fire in the Lake: The Vietnamese and the Americans in Vietnam* (New York: Atlantic Monthly Press, 1972), p. 351.

204   "held its own dangers and terrors"   John D. Bergen, *Military Communications: A Test for Technology* (Washington, D.C.: Center of Military History, 1986), p. 340.

204   the invention of maps   See G. Malcolm Lewis, "The Origins of Cartography," in *The History of Cartography,* vol. I: Cartography in Prehistoric, Ancient, and Medieval Europe and the Mediterranean, ed. J. B. Harley and David Woodward (Chicago: University of Chicago Press, 1987), pp. 50–53.

206   "trying all of the apartment doors"   Police records, Ridgefield Park, N.J.

207   "a land rich in gold"   Quoted in Stephen Clissold, *The Seven Cities of Cíbola* (London: Eyre & Spottiswoode, 1961), pp. 101–102.

208   "It must exist"   Bernard DeVoto, *The Course of Empire* (Boston: Houghton Mifflin, 1952), p. 63.

208   "it had become . . . a mighty stream"   Carl I. Wheat, *Mapping the Transmississippi West, 1540–1861,* vol. 1: The Spanish Entrada to the Louisiana Purchase, 1540–1804 (San Francisco: Institute of Historical Cartography, 1957), pp. 140–141.

209   "the existence of a great river"   Frémont, *Report of the Expedition to the Rockies,* p. 196.

209   "with every stream . . . to see the great Buenaventura"   Ibid., p. 219.

209   "appallingly foolhardy"   Nevins, *Frémont,* p. 168.

211   "nervous disorder"   Jail docket, Bergen County Sheriff's Office, July 6, 1971.

211   "denie[d] illnesses"   Jail docket, Bergen County Sheriff's Office, January 21, 1971.

212   "going crazy"   Gilbert Bland, letter to Gov. William T. Cahill, September 29, 1972. Records of Gov. William T. Cahill, New Jersey State Archives.

212   a warrant for his arrest   Bench warrant issued in Sussex County, N.J., June 22, 1973.

213   "give consideration to retaining legal counsel"   Gov. William T. Cahill, letter to Gilbert Bland, October 5, 1972. Cahill mistakenly spelled the name *Blond*.

213   " 'snitching' as they call it"   Gilbert Bland, letter to U.S. District Judge Ben Krentzman, May 10, 1976.

214   "I have no intention"   Gilbert Bland, letter of unknown date to U.S. District Judge Ben Krentzman; received by Krentzman, April 6, 1977.

215   "Frémont's march south"   John Noble Wilford, *The Mapmakers* (New York: Alfred A. Knopf, 1981), p. 199.

215   "no hesitation in asserting"   *The Journals of Zebulon Montgomery Pike: With Letters and Related Documents,* vol. 2, ed. Donald Jackson (Norman: University of Oklahoma Press, 1966), pp. 26–27.

Chapter Nine: The Waters of Paradise

217   "a spring of running water"   Quoted in Samuel Eliot Morison, *The European Discovery of America: The Southern Voyages,* A.D. *1492–1616* (New York: Oxford University Press, 1974), p. 504.

217   "noble and beautiful well"   *The Travels of Sir John Mandeville,* trans. C.W.R.D. Moseley (Harmondsworth, England: Penguin Books, 1983), p. 123.

218   *el enflaquecimiento del sexo*   See Morison, *European Discovery of America,* p. 503.

218   lowest proportions of locally born citizens   In the 1990 U.S. Census, Florida ranked dead last for its percentage of residents

born in-state. And, according to Census Bureau statistics, Florida was the tenth fastest-growing state in the country between 1990 and 1998.

219   "His interest in maps"   James R. Creekmore, letter to author, October 30, 1996. This letter included a "Statement Concerning Mr. Gilbert Bland," responding to a detailed list of questions I had sent to the map thief through his attorneys. Although the statement was only four paragraphs long—less than one full page—and contained almost no direct answers, it was the only response Bland would ever make to my inquiries about his life and motives.

219   owned by Karen Bland   Records of Maryland Department of Assessment and Taxation, Corporate Division.

219   "major customers"   Paul R. Thomson, Jr., and James R. Creekmore, "Defendant's Objections to Presentence Investigation Report," *United States of America vs. Gilbert Joseph Bland*, Charlottesville, Va., September 16, 1996.

220   "no longer . . . economically feasible"   Transcript of sentencing hearing in *United States of America vs. Gilbert Joseph Bland*, U.S. District Court, Charlottesville, Va., December 9, 1996.

221   "everyplace looks like noplace"   James Howard Kunstler, *The Geography of Nowhere: The Rise and Decline of America's Man-Made Landscape* (New York: Simon & Schuster, 1993), p. 131.

232   "For God doth know"   Genesis 3:5.

232   "eastward in Eden"   Genesis 2:8.

232   an orchard surrounded by a wall   See Jean Delumeau, *History of Paradise: The Garden of Eden in Myth and Tradition*, trans. Matthew O'Connell (New York: Continuum, 1995), p. 4.

233   "wall of fire"   Quoted in ibid., p. 44.

233   "great evidence of the earthly Paradise"   Christopher Columbus, "Narrative of the Third Voyage," in *The Four Voyages of Christopher Columbus*, trans. J. M. Cohen (New York: Penguin, 1969), pp. 220–222.

233   "the flood, and other accidents of time"   Walter Raleigh, *The History of the World,* book one (1614; reprint, London: printed for G. Conyers, et al., 1736), p. 23.

234   "shaped . . . when opened"   Quoted in Simon Berthon and Andrew Robinson, *The Shape of the World* (London: George Philip, 1991), pp. 57–58.

235   "See you later"   See Frank D. Roylance's account, "Fla. Man Is Sought in Book Cuttings: He and Wife Abruptly Close Map Shop," in Baltimore *Sun* (December 16, 1995): 1A.

236   declared Chapter Seven bankruptcy   Papers filed in U.S. Bankruptcy Court, Southern District of Florida, Fort Lauderdale Division, November 3, 1995.

CHAPTER TEN: THE JOY OF DISCOVERY

239   "I discovered"   Christopher Columbus, "The Barcelona Letter of 1493," trans. Lucia Graves, in Mauricio Obregón, ed., *The Columbus Papers: The Barcelona Letter of 1493, the Landfall Controversy, and the Indian Guides* (New York: Macmillan, 1991), pp. 65–69. All further quotations from the letter come from the Graves translation, with one exception: I use Felipe Fernández-Armesto's choice of wording for the phrase "the conquest of what appears impossible."

241   *"He was charmed"*   Gianni Granzotto, *Christopher Columbus,* trans. Stephen Sartarelli (Garden City, N.Y.: Doubleday, 1985), p. 34.

244   OLD MAP POX   This poster was created by the map dealer F. J. Manasek.

252   "the first map of the whole world"   Peter Whitfield, *The Image of the World: Twenty Centuries of World Maps* (San Francisco: Pomegranate Artbooks, 1994), p. 50.

253   "chanced to notice a small copper globe"   Henry Stevens, *Recollections of Mr. James Lenox of New York and the Formation of His Library* (London: Henry Stevens & Son, 1886), p. 141.

254   "the final thrill"   Walter Benjamin, "Unpacking My Library:

A Talk About Book Collecting," in *Illuminations,* trans. Harry Zohn, ed. Hannah Arendt (New York: Harcourt, Brace & World, 1968), p. 60.

255   "I have not a hair upon me"   Christopher Columbus, "Letter of Columbus to Their Majesties," in *Select Documents Illustrating the Four Voyages of Christopher Columbus,* trans. and ed. Cecil Jane (London: Hakluyt Society, 1933), pp. 108–110. The other Columbus quotations in this paragraph are from the same letter.

255   *"Almost anyone . . . would rest content"*   Felipe Fernández-Armesto, *Columbus* (New York: Oxford University Press, 1991), p. 192.

257   "search—successful or not"   Werner Muensterberger, *Collecting: An Unruly Passion* (Princeton, N.J.: Princeton University Press, 1994), p. 162.

258   "There is reason to believe"   Ibid., pp. 10, 13.

259   "The quest is never-ending"   Ibid., p. 162.

259   "an all-consuming passion"   Ibid., p. 6.

260   the notorious Don Vincente   See Nicholas A. Basbanes, *A Gentle Madness: Bibliophiles, Bibliomanes, and the Eternal Passion for Books* (New York: Henry Holt, 1995), pp. 33–34.

262   the story of the Unknown Pilot   See Samuel Eliot Morison's overview in *Admiral of the Ocean Sea: A Life of Christopher Columbus* (Boston: Little, Brown, 1942), pp. 61–63, as well as Gronzotto's account in *Christopher Columbus,* pp. 45–47.

262   "a protective figure"   Joseph Campbell, *The Hero with a Thousand Faces,* 2d ed. (Princeton, N.J.: Princeton University Press, 1968), pp. 69, 71.

263   "power-imbued fetish[es]"   Muensterberger, *Collecting,* p. 54.

263   "set [his] mind ablaze"   Quoted in Granzotto, *Christopher Columbus,* p. 57.

263   "still more inflamed"   Quoted in John Noble Wilford, *The Mysterious History of Columbus: An Exploration of the Man, the Myth, the Legacy* (New York: Alfred A. Knopf, 1991), p. 76.

263   "Some, to beautify their Halls"   Quoted in John Goss, *The*

Mapmaker's Art: An Illustrated History of Cartography (Skokie, Ill.: Rand McNally, 1993), p. 344.

265 "As Freud has shown" Campbell, *Hero with a Thousand Faces*, p. 51.

270 *"Something hidden. Go and find it"* Rudyard Kipling, "The Explorer," in *Complete Verse* (London: Kyle Cathie, 1990), pp. 86–88.

CHAPTER ELEVEN: THE ISLAND OF LOST MAPS

273 El Wakwak See Raymond H. Ramsay, *No Longer on the Map: Discovering Places That Never Were* (New York: Viking Press, 1972), p. 113.

273 "sirens abound" Quoted in William G. Niederland's essay "The Pre-Renaissance Image of the World and the Discovery of America," in *Maps from the Mind: Readings in Psychogeography*, ed. Howard F. Stein and William G. Niederland (Norman: University of Oklahoma Press, 1989), p. 111.

274 "men who murdered their fathers" Quoted in ibid., p. 111.

274 "produces all crops and all fruits" Quoted in Jean Delumeau, *History of Paradise: The Garden of Eden in Myth and Tradition*, trans. Matthew O'Connell (New York: Continuum, 1995), p. 100.

274 "there are trees that yield birds" Quoted in ibid., p. 103.

274 "have tails like animals" Quoted in E. G. Ravenstein, *Martin Behaim: His Life and His Globe* (London: George Philip & Son, 1908), p. 88.

274 "One of these islands [is] inhabited by men only" Quoted in ibid., p. 105.

274 Drogeo The Drogeo and Icaria legends stem from Nicolò Zeno's apocryphal sixteenth-century account of his ancestors' purported adventures. See Nicolò Zeno, *The Voyages of the Venetian Brothers, Nicolò & Antonio Zeno, to the Northern Seas in the XIVth Century*, ed. and trans. Richard Henry Major (London: Hakluyt Society, 1873), pp. 21–34. See also William H.

Babcock, *Legendary Islands of the Atlantic: A Study in Medieval Geography* (New York: American Geographical Society, 1922), pp. 124–142, as well as Ramsay, *No Longer on the Map,* pp. 53–76. Ramsay also covers the St. Brendan legend, pp. 77–85.

275   "full of men agrieving and lamenting"   Quoted in Howard Rollin Patch, *The Other World: According to the Descriptions in Medieval Literature* (Cambridge, Mass.: Harvard University Press, 1950), p. 35. In addition to Patch, information on Celtic island lore comes from Donald S. Johnson, *Phantom Islands of the Atlantic: The Legends of Seven Lands That Never Were,* rev. ed. (New York: Walker, 1996), Ramsay, *No Longer on the Map,* and John Kirtland Wright, *The Geographical Lore of the Time of the Crusades: A Study in the History of Medieval Science and Tradition in Western Europe* (New York: American Geographical Society, 1925).

275   "a bird of monstrous size"   *Tales from the Thousand and One Nights,* trans. N. J. Dawood (New York: Penguin, 1973), p. 124.

275   Island of Jewels   See Joseph Campbell, *The Hero with a Thousand Faces,* 2d ed. (Princeton, N.J.: Princeton University Press, 1968), pp. 114–115.

275   "full of noises"   William Shakespeare, *The Tempest,* act III, sc. 2, lines 133–134.

275   the tale of Maushope   See Seon and Robert Manley's description in *Islands: Their Lives, Legends, and Lore* (Philadelphia: Chilton, 1970), pp. 18–21.

275   "where all the four seasons"   J. M. Barrie, *Peter Pan,* act II, sc. 1, lines 37–38.

275   "way west of Sumatra"   Dialogue from *King Kong,* screenplay by James Creelman and Ruth Rose, 1933.

276   "forested slopes were wreathed"   Michael Crichton, *Jurassic Park* (New York: Alfred A. Knopf, 1990), p. 79.

277   "misery acquaints a man"   Shakespeare, *Tempest,* act II, sc. 2, lines 40–41.

277   "a soul or a life-force"   Werner Muensterberger, *Collecting: An*

Unruly Passion (Princeton, N.J.: Princeton University Press, 1994), p. 55.

277   "the relationship between collector and dealer"   Ibid., p. 159.

282   as much as half a million dollars   No official dollar estimate for the maps exists. This is a best-guess figure, based on my conversations with several map experts and law enforcement officials familiar with the case.

282   "recovered, unharmed"   News release, U.S. Attorney's Office, Western District of Virginia, March 1, 1996.

285   "all possibility of deliverance"   Daniel Defoe, *Robinson Crusoe,* ed. Angus Ross (1719; reprint, Harmondsworth, England: Penguin, 1965), p. 112.

287   "wild ideas"   Ibid., p. 162. Further Defoe quotations in this chapter also come from the Penguin edition, pp. 162–165.

Chapter Twelve: Eldorado

289   Great Indian Trading Path   See Douglas L. Rights, "The Trading Path to the Indians," *North Carolina Historical Review,* vol. 8, no. 4 (October 1931): 403–426. See also Douglas Summers Brown's chapter on the path in *The Catawba Indians: The People of the River* (Columbia: University of South Carolina Press, 1966), pp. 69–100.

291   an Indian village   Some experts have argued that the forerunner of present-day Hillsborough is marked "Oenock" on the Ogilby-Moxon map. But R. P. Stephen Davis, Jr., a University of North Carolina research archaeologist who is excavating Indian settlements near downtown Hillsborough, told me that he believes the site is more likely to be a place Ogilby recorded as "Sabor."

291   "covetous and thievish"   John Lederer, "The First Expedition, From the head of the Pemaeoncock, alias York-River (due West) to the top of the Apalataean Mountains," in *The Discov-*

eries of John Lederer, ed. William P. Cumming (Charlottesville: University of Virginia Press, 1958), p. 27.

293   "the great and Golden Citie"   This phrase comes from the subtitle of Raleigh's book, *The Discoverie of the Large, Rich and Bewtiful Empyre of Guiana, with a relation of the great and Golden Citie of Manoa (which the Spanyards call El Dorado),* published in 1596.

293   the first explorer known to have put Manoa   See Raymond H. Ramsay's discussion in *No Longer on the Map: Discovering Places That Never Were* (New York: Viking Press, 1972), pp. 17–18.

294   "a man of many characters"   Robert Silverberg, *The Golden Dream: Seekers of El Dorado* (Athens: Ohio University Press, 1996), p. 307.

294   "There are many possible readings"   Charles Nicholl, *The Creature in the Map: A Journey to El Dorado* (New York: William Morrow, 1995), p. 23.

294   "the graves have not beene opened"   Walter Raleigh, *The Discoverie of the Large, Rich and Bewtiful Empyre of Guiana,* ed. Neil L. Whitehead (Norman: University of Oklahoma Press, 1997), p. 196.

295   "the deare crossing in every path"   Ibid., p. 176.

297   80 percent in fifty years   See Mark Warhus's discussion in *Another America: Native American Maps and the History of Our Land* (New York: St. Martin's Press, 1997), p. 80.

298   "Indigenous Americans communicated information"   Mark Monmonier, *Drawing the Line: Tales of Maps and Cartocontroversy* (New York: Henry Holt, 1995), p. 107.

298   renaming perhaps half of the seven hundred thousand towns and cities   See Derek Nelson's discussion in *Off the Map: The Curious History of Place Names* (New York: Kodansha International, 1997), p. 113.

300   "there is only the journey"   Nicholl, *Creature in the Map,* p. 11.

300   like so many dreamers, desperadoes, and idiots   Much of my background material for this paragraph comes from *The Search for El Dorado,* series ed. Dale M. Brown (Alexandria, Va.: Time-Life Books, 1994), pp. 7–34.

301   "The reports are false"   Quoted in Nicholl, *Creature in the Map,* p. 28.

301   "gallant knight"   Edgar Allan Poe, "Eldorado," in *Edgar Allan Poe: Sixty-seven Tales* (New York: Gramercy Books, 1990), p. 752.

CHAPTER THIRTEEN: MR. BLAND, I PRESUME

305   "I have taken a solemn, enduring oath"   Henry Morton Stanley, *How I Found Livingstone: Travels, Adventures, and Discoveries in Central Africa* (New York: Charles Scribner's Sons, 1913), pp. 308–309.

306   "fevers without number"   Ibid., p. 309.

307   "no theft nor sycophancy"   Letter quoted in Jean Delumeau, *History of Paradise: The Garden of Eden in Myth and Tradition,* trans. Matthew O'Connell (New York: Continuum, 1995), p. 76. This letter—written to the Byzantine emperor by someone claiming to be Prester John, around 1165—was widely circulated. "No one in the Middle Ages doubted its authenticity," wrote Delumeau.

307   "the Lost-Boy Complex"   See Eric Leed's chapter on the subject in *Shores of Discovery: How Expeditionaries Have Constructed the World* (New York: Basic Books, 1995), pp. 198–230.

307   more than twenty expeditions   It is interesting that the man who backed perhaps the most famous of these expeditions was George Peabody, founder of that great library in Baltimore. According to Leed, the philanthropist underwrote Elisha Kent Kane's 1853 search for Franklin. A bay off the north of Greenland was later named Peabody Bay.

307   "The Franklin affair was the greatest disaster"   Peter Whit-

field, *New Found Lands: Maps in the History of Exploration* (New York: Routledge, 1998), p. 181.

308 "I . . . begin to think myself somebody" Quoted in John Bierman, *Dark Safari: The Life Behind the Legend of Henry Morton Stanley* (New York: Alfred A. Knopf, 1990), p. 118.

309 Dutch master Jan Vermeer See James A. Welu, "Vermeer: His Cartographic Sources," *Art Bulletin,* vol. 57, no. 4 (December 1975): 529–547.

309 *"In thy face I see"* William Shakespeare, *Henry VI, Part II,* act III, sc. 1, lines 202–203.

310 "I found myself gazing at him" Stanley, *How I Found Livingstone,* p. 413.

310 "vent [his] joy" Ibid., p. 411.

311 had trouble concentrating Gilbert Bland, letter to U.S. District Court Judge James H. Michael, Jr., April 16, 1996.

312 "pattern of problems" Transcript of sentencing hearing in *United States of America vs. Gilbert Joseph Bland,* U.S. District Court, Charlottesville, Va., December 9, 1996.

312 "The first thing I'd like to say" Ibid.

313 "In the court's view" Ibid.

313 "a persistent misunderstanding" Order by Judge Michael, January 8, 1997.

314 "the penalty is not severe enough" Quoted in Todd Nelson, "Judge Vetoes Plea Bargain in Map Case: The Plea Arrangement for a Man Who Admitted Taking Papers from UNC-CH Is Termed Too Lenient," Raleigh, N.C., *News & Observer* (December 20, 1996): 1B.

322 "I think the biggest punishment" Paul R. Thomson, Jr., in transcript of sentencing hearing in *United States of America vs. Gilbert Joseph Bland,* December 9, 1996.

322 "To discover is to draw the veil" Mauricio Obregón, "Why Columbus?" in *The Columbus Papers: The Barcelona Letter of 1493, the Landfall Controversy, and the Indian Guides,* ed. Mauricio Obregón (New York: Macmillan, 1991), p. 3.

323   literally millions of people   See Adam Hochschild, *King Leopold's Ghost: A Story of Greed, Terror, and Heroism in Colonial Africa* (Boston: Houghton Mifflin, 1998), pp. 225–234. Hochschild estimates that through murder, starvation, exhaustion, exposure, disease, and declining birthrates, the population in the Congo Free State declined by 10 million as a result of Leopold's brutal policies.

323   "a genuine interest"   James R. Creekmore, letter to author, October 30, 1996.

325   "Nature conceives of innumerable things"   Quoted in Peter Whitfield, *The Image of the World: Twenty Centuries of World Maps* (San Francisco: Pomegranate Artbooks, 1994), p. 2.

EPILOGUE: LIFTING OFF

328   "I can see the whole state"   Quoted in John Noble Wilford, *The Mapmakers* (New York: Alfred A. Knopf, 1981), p. ix.

332   "The authorities never contacted me"   Although this source did not request anonymity—and although he was well aware that I was a journalist working on the Bland story—I have chosen not to mention his name, since his comments did not occur within the context of a formal interview.

341   "I'm under [psychiatric] medicine"   University of Delaware videotape of Gilbert Bland, March 27, 1997.

344   The most telling picture I ever saw   See Corey Lowenstein's photo accompanying Jay Price's story, "Quiet Appearance for Map Thief," Raleigh, N.C., *News & Observer* (July 2, 1996): 1B.

345   "the world's most powerful civilian spacecraft"   William J. Broad, "Private Spy in Space to Rival Military's," *New York Times* (April 27, 1999): D1–2.

345   "We find ourselves in the midst"   Stephen S. Hall, *Mapping the Next Millennium: The Discovery of New Geographies* (New York: Random House, 1992), p. 6.

345   "It's like being able to drain"   The oceanographer David
      Sandwell, quoted in Ken Miller, "Spy Data Gives First Clear
      View of Sea Floor," Gannet News Service (October 23, 1995).
      See also Earl Lane, "Bottom of Sea Viewed in Detail: Navy
      Satellite Helps Map Oceans' Floors," *Newsday* (October 24,
      1995): A17.

345   Sloan Digital Sky Survey   See Kelly Campbell, "What's Next:
      A 3-D Map of 1 Billion Light Years," *Newsday* (September 29,
      1998): C13; Charles W. Petit, "Mapping the Heavens," *U.S.
      News & World Report*, vol. 124, no. 22 (June 22, 1998): 56; and
      Alexandra Witze, "The Big Picture," *Dallas Morning News*
      (April 27, 1998): 6D.

346   create life-forms   See Roger Highfield, "We Can Create
      Life, Says Scientist," *Daily Telegraph* of London (January 25,
      1999).

347   "opening up the territory of the mind"   Rita Carter, *Mapping
      the Mind* (Berkeley and Los Angeles: University of California
      Press, 1998), p. 6.

347   "an individual's state of mind"   Ibid., pp. 6–7.

348   "These are the early days"   Ibid., p. 8.

349   They never found it   Although investigators never located the
      satellite, they believe that they know what happened to it. In
      reviewing the mishap, a panel of experts concluded that, be-
      cause of an electrical problem, the aerodynamic covering over
      the payload failed to come off properly. Because the covering
      did not fall away from the rocket, the satellite was probably
      dragged back down into the atmosphere, where it burned up.
      Another *Ikonos* satellite was launched—successfully—on Sep-
      tember 24, 1999.

Index

Page numbers in italics refer to illustrations.

/

Excerpts from *The Story of Maps*, by Lloyd Brown (Little, Brown & Company, 1949). Copyright © 1949 by Lloyd Arnold Brown. Copyright renewed 1977 by Florence D. Brown.

Grateful acknowledgment is made to the following for permission to reprint unpublished and previously published material:

The Baltimore Sun: "Auction of Peabody Books Draws Approval, Outrage," by John Dorsey, Baltimore *Sun*. Reprinted by permission of the Baltimore *Sun*.

Doubleday, a division of Random House, Inc., and Macmillan General Books: Excerpt from *Worldly Goods: A New History of the Renaissance*, by Lisa Jardine. Copyright © 1996 by Lisa Jardine. Rights in the British Commonwealth are controlled by Macmillan General Books, London. Reprinted by permission of Doubleday, a division of Random House, Inc., and Macmillan General Books.

Daniel Mark Epstein: Excerpt from "Mr. Peabody & His Athenaeum," by Daniel Mark Epstein. Copyright © 1995 by Daniel Mark Epstein. All rights reserved. Reprinted by permission of Daniel Mark Epstein.

Johns Hopkins University: An e-mail dated December 7, 1995, posted to the Ex Libris newsgroup by Cynthia Requardt, entitled "Theft." All rights reserved. Reprinted by permission of Johns Hopkins University.

F. J. Manasek: Text from "map pox" poster, courtesy of F. J. Manasek. Used by permission.

Random House, Inc.: Excerpt from *The Discoverers*, by Daniel J. Boorstin. Copyright © 1983 by Daniel J. Boorstin. Reprinted by permission of Random House, Inc.

Tribune Media Services: Article entitled "Police on the Trail of a Map Thief." Reprinted with permission of Knight Ridder/Tribune Information Services.

The author and publisher wish to thank the following institutions for photographs and photographic reproductions: Edward E. Ayer Collection, The Newberry Library, Chicago (pp. x, 44, 66, 111, 194, 216, 240, 272, 296); Maryland Historical Society, Baltimore, Maryland (p. 2); Archives of the Peabody Institute of the Johns Hopkins University (p. 106); William L. Clements Library, University of Michigan (pp. 134, 290); *The News & Observer*/Corey Lowenstein (p. 304).

MILES HARVEY began reporting on Gilbert Bland in 1996 for *Outside* magazine. He has worked for UPI and *In These Times,* and he was the book-review columnist for *Outside.* A graduate of the University of Illinois at Champaign-Urbana and the University of Michigan, he has had a lifelong fascination with maps. He can be reached via the Internet at www.milesharvey.com.